T0137151

Sustainable Civil Infrastructures

Editor-in-Chief

Hany Farouk Shehata, SSIGE, Soil-Interaction Group in Egypt SSIGE, Cairo, Egypt

Advisory Editors

Khalid M. ElZahaby, Housing and Building National Research Center, Giza, Egypt
Dar Hao Chen, Austin, TX, USA

Sustainable Infrastructure impacts our well-being and day-to-day lives. The infrastructures we are building today will shape our lives tomorrow. The complex and diverse nature of the impacts due to weather extremes on transportation and civil infrastructures can be seen in our roadways, bridges, and buildings. Extreme summer temperatures, droughts, flash floods, and rising numbers of freeze-thaw cycles pose challenges for civil infrastructure and can endanger public safety. We constantly hear how civil infrastructures need constant attention, preservation, and upgrading. Such improvements and developments would obviously benefit from our desired book series that provide sustainable engineering materials and designs. The economic impact is huge and much research has been conducted worldwide. The future holds many opportunities, not only for researchers in a given country, but also for the worldwide field engineers who apply and implement these technologies. We believe that no approach can succeed if it does not unite the efforts of various engineering disciplines from all over the world under one umbrella to offer a beacon of modern solutions to the global infrastructure. Experts from the various engineering disciplines around the globe will participate in this series, including: Geotechnical, Geological, Geoscience, Petroleum, Structural, Transportation, Bridge, Infrastructure, Energy, Architectural, Chemical and Materials, and other related Engineering disciplines.

More information about this series at http://www.springer.com/series/15140

Hany Shehata · Heinz Brandl ·
Mounir Bouassida · Tamer Sorour
Editors

Sustainable Thoughts in Ground Improvement and Soil Stability

Proceedings of the 3rd GeoMEast
International Congress and Exhibition, Egypt
2019 on Sustainable Civil Infrastructures –
The Official International Congress
of the Soil-Structure Interaction Group
in Egypt (SSIGE)

 Springer

Editors
Hany Shehata
Soil Structure Interaction Group Egypt
Cairo, Egypt

Heinz Brandl
Vienna University of Technology
Vienna, Austria

Mounir Bouassida
National Engineering School of Tunis
(ENIT)
Tunis, Tunisia

Tamer Sorour
Ain-Sham University
Cairo, Egypt

ISSN 2366-3405   ISSN 2366-3413   (electronic)
Sustainable Civil Infrastructures
ISBN 978-3-030-34183-1   ISBN 978-3-030-34184-8   (eBook)
https://doi.org/10.1007/978-3-030-34184-8

This Springer imprint is published by the registered company Springer Nature Switzerland AG
The registered company address is: Gewerbestrasse 11, 6330 Cham, Switzerland

Contents

About the Editors

Hany Shehata is the founder and CEO of the Soil-Structure Interaction Group in Egypt "SSIGE". He is a partner and Vice-President of EHE-Consulting Group in the Middle East and managing editor of the "Innovative Infrastructure Solutions" journal, published by Springer. He worked in the field of civil engineering early, while studying, with Bechtel Egypt Contracting & PM Company, LLC. His professional experience includes working in culverts, small tunnels, pipe installation, earth reinforcement, soil stabilization and small bridges. He also has been involved in teaching, research and consulting. His areas of specialization include static and dynamic soil-structure interactions involving buildings, roads, water structures, retaining walls, earth reinforcement and bridges, as well as, different disciplines of project management and contract administration. He is the author of an Arabic practical book titled "Practical Solutions for Different Geotechnical Works: The Practical Engineers' Guidelines". He is currently working on a new book titled "Soil-Foundation-Superstructure Interaction: Structural Integration". He is the contributor of more than 50 publications in national and international conferences and journals. He served as a co-chair of the GeoChina 2016 International Conference in Shandong, China. He serves also as a co-chair and secretary general of the GeoMEast 2017 International Conference in Sharm El-Sheikh, Egypt. He received the Outstanding Reviewer of the ASCE for 2016 as selected by the Editorial Board of International Journal of Geomechanics.

Prof. Heinz Brandl has been Full Professor for Soil and Rock Mechanics and Foundation Engineering (including Tunnelling) since 1977, chairing until 2009 the prestigious Institute for Soil Mechanics and Geotechnical Engineering, which was founded by Prof. Karl Terzaghi in 1928 at the Vienna University of Technology. Since 2008, he is Emeritus Professor.

Professor Brandl authored about 580 scientific publications (including 21 books), partly published in 18 languages. The subjects cover laboratory and field testing, soil and rock mechanics, foundation engineering, slope engineering, earthworks, tunnelling, urban undergrounds, restoration of historical buildings, road and railway engineering, hydro/hydraulic engineering and environmental engineering, geosynthetics, geothermal engineering ("energy foundations", "energy tunnels", etc.), natural disaster mitigation and rehabilitation, etc. He also published on philosophical aspects and on ethics in the profession. From the very outset of his professional work, H. Brandl has been bridging the gap between theory and practice. He has been fully responsible for nearly 4000 projects of civil engineering, geotechnical and environmental engineering in Austria and elsewhere.

Prof. Brandl has been active worldwide since 1968 as chairman, general reporter, state-of-the-art reporter, special-, keynote- and opening lecturer, discussion leader and panellist at numerous international conferences on geotechnical engineering, environmental engineering, geosynthetics, etc. He was Rankine Lecturer (2001), Giroud Lecturer (2010), and he created the prestigious "Vienna Terzaghi Lecture". He was Vice-President, ISSMGE (1997–2001), and from 1973 to 2015 President of the Austrian Geotechnical Society. Moreover, he is a member of the Royal Academy of Sciences of Belgium, of the International Academy of Engineers (Moscow) and other Scientific Academies. Since June 2003, he has been President of the Austrian Society for Engineers and Architects (founded in 1848).

He received numerous national and international awards (e.g. Kevin Nash Gold Medal), honorary doctorates and other honours.

Mounir Bouassida is a professor of civil engineering at the National Engineering School of Tunis (ENIT) of the University of Tunis El Manar where he earned his B.S., M.S., Ph.D., and doctorate of sciences diplomas, all in civil engineering. He is the director of the Research Laboratory in Geotechnical Engineering and Georisk. Dr. M. Bouassida has supervised 20 Ph.D. and 30 master of science graduates. His research focuses on soil improvement techniques and behaviour of soft clays. Dr. Bouassida is the (co)author of about 91 papers in refereed international journals; 130 papers, including 21 keynote lectures; and three books. He is also co-editor of ten published proceedings of international conferences.

He is a member of the editorial committees of journals Ground Improvement (ICE), Geotechnical Geological Engineering, Infrastructure Innovative Solutions and International Journal of Geomechanics (ASCE). He is also an active reviewer in several international journals. As a 2006 Fulbright scholar, Dr. Bouassida elaborated a novel methodology for the design of foundations on reinforced soil by columns. He was awarded the 2006 S. Prakash Prize for Excellence in the practice of geotechnical engineering. In 2008, Bouassida launched a Tunisian consulting office in geotechnical engineering, SIMPRO. He is a co-developer of the software Columns 1.01 used for designing column-reinforced foundations. Prof. Bouassida held the office of the Vice-President of ISSMGE for Africa (2005–2009). He benefited from several grants as a visiting professor in the USA, France, Belgium, Australia, Vietnam, Hong Kong and Norway. Recently, Prof. Bouassida became an appointed member of the ISSMGE board (2017–2021). He is managing the webinars activity within the newly established programme of ISSMGE virtual university.

Assistant Professor Tamer Sorour, Faculty of Engineering, Ain Shams University, Cairo, Egypt
Assistant Professor Tamer Sorour is a Geotechnical Engineer graduated from the Faculty of Engineering, Ain Shams University, and obtained his PhD from the Faculty of Engineering, Ain Shams University in 2009. He performed various researches in different fields of Geotechnical Engineering, such as soil dynamics including the hazards accompanied with pile driving in different soil formations as well as the design of machine foundations and soil liquefaction and the stability of soil slopes under the effect of earthquakes. The field of deep excavations including different shoring and anchorage systems. The field of soil improvement using stone columns and the effect of the installation technique on the behaviour of the stone columns. He also shared in different researches that are focused on the response of large diameter piles in different soil formations. He also have minor/major contribution in different mega projects in Egypt as supervisor of structural and geotechnical works including, but not limited to: geotechnical and geo-environmental studies, execution of boreholes, shallow and deep excavation works, performing different laboratory testing, field control and field testing using different techniques, installation of different types of piles, retaining walls and different supported deep excavation systems, soil stabilization, dewatering, protection of side slopes and landslides, grouting and field monitoring.

Determination of Shear-Wave Velocity to 30 m Depth from Refraction Microtremor Arrays (Remi-Test). Applied in Some Greek-Roman Archaeological Sites in Alexandria, Egypt

Sayed Hemeda[✉]

Conservation Department, Faculty of Archaeology, Cairo University, Giza,
Egypt
sayed.hemeda@cu.edu.eg

Abstract. Preservation of monuments all around the world and increasing their stability against earthquakes is a matter of great importance. Main purpose of the present study is to investigate the dynamic characteristics of the underground monuments at the ancient sites in Alexandria, Egypt (Catacomb of Kom El-Shoqafa, El-Shatbi Necropolis, and Necropolis of Mustafa Kamil, Amd El-Sawari site (Serapium and ancient library)) and identify the main damage mechanism, in order to evaluate the risk of structure damage or collapse in case of future events using microtremors recordings. Array measurements at three sites in the city of Alexandria were performed to estimate the Vs velocity of soil/rock formations for site effect analysis. Our study includes a detailed geological and geotechnical survey of the areas, measurement, analysis and interpretation of ambient noise data using the refraction microtremor (ReMi) method.

A thorough assessment of shallow shear velocity is important to both earthquake-hazard assessment and efficient foundation design. The only standard procedure for determining shear velocity, crosshole seismic (ASTM D4428), requires at least two boreholes with high-precision positional logs. The refraction microtremors method is based on recording ambient ground noise on simple seismic refraction equipment (as in ASTM D5777). Wave field analysis of the noise allows picking of Rayleigh-wave phase velocities. It works well in dense urban areas and transportation corridors.

The shear velocities estimated from ReMi method is a fast commercial effective method as borehole velocities for estimating 30- m depth-averaged shear velocity for foundation design and other purposes.

The importance of soil shear wave velocity (Vs) can not be over emphasized in engineering work. Because of its vital importance, many field and lab techniques have been devised to obtain soil Vs including, SPT, CPT, SCPT, P-S logging, suspension logging, cross-hole, seismic refraction and reflection (e.g. Imai 1981; Japan Road Association 1990; AIJ 1993; Kramer 1996). All of these techniques are inaccurate, costly, intrusive (require boring), laborious, time-consuming, or not urban-friendly.

Current commonly used techniques of estimating shallow shear velocities for assessment of earthquake site response are too costly for use in most urban areas. They require large sources to be effective in noisy urban settings, or specialized independent recorders laid out in an extensive array. The refraction

© Springer Nature Switzerland AG 2020
H. Shehata et al. (Eds.): GeoMEast 2019, SUCI, pp. 1–34, 2020.
https://doi.org/10.1007/978-3-030-34184-8_1

microtremor (ReMi) method overcomes these problems by using standard P-wave recording equipment and ambient noise to produce average one-dimensional shear-wave profiles down to 100 m depths. The combination of commonly available equipment, simple recording with no source, a wave field transformation data processing technique, and an interactive Rayleigh-wave dispersion modeling tool exploits the most effective aspects of the microtremor, spectral analysis of surface wave (SASW), and multichannel analysis of surface wave (MASW) techniques. The slowness-frequency wave field transformation is particularly effective in allowing accurate picking of Rayleigh-wave phase-velocity dispersion curves despite the presence of waves propagating across the linear array at high apparent velocities, higher-mode Rayleigh waves, body waves, air waves, and incoherent noise. It has been very effective for quickly and cheaply determining 30-m average shear wave-velocity (V30).

Use of "active source" methods such as seismic reflection and refraction. In geotechnical applications in particular, seismic refraction with surface seismic sources has gained widespread acceptance as a viable investigation tool (Whiteley 1994). The effectiveness of this approach, especially in urban situations, is limited by the presence of seismic noise and in the choice of a source with sufficient energy to achieve the required depth penetration. Additionally, the seismic refraction method is inherently "blind" to the presence of a velocity inversion (Whiteley and Greenhalgh 1979).

An alternative approach is to use "natural" microtremors (the "noise" in traditional seismic surveying), as a source of wave energy. The measurement of high-frequency seismic noise, or microtremors, is a well-established method of estimating the seismic resonance characteristics of relatively thick (tens of metres and above) unconsolidated sediments. This approach is described by Nakamura (1989), where the fundamental resonance period (TS) of a site can be obtained from surface waves and used in the assessment of potential seismic hazard to structures founded in soft soils.

The refraction microtremor technique is based on two fundamental ideas. The first is that common seismic refraction recording equipment, set out in a way almost identical to shallow P-wave refraction surveys, can effectively record surface waves at frequencies as low as 2 Hz. The second idea is that a simple, two-dimensional slowness-frequency (p-f) transform of a microtremor record can separate Rayleigh waves from other seismic arrivals, and allow recognition of true phase velocity against apparent velocities.

Two essential factors that allow exploration equipment to record surface-wave velocity dispersion, with a minimum of field effort, are the use of a single geophone sensor at each channel, rather than a geophone "group array," and the use of a linear spread of 12 or more geophone sensor channels. Single geophones are the most commonly available type, and are typically used for refraction rather than reflection surveying. The advantages of ReMi from a seismic surveying point of view are several, including the following: It requires only standard refraction equipment already owned by most consultants and universities; it requires no triggered source of wave energy; and it will work best in a seismically noisy urban setting. Traffic and other vehicles, and possibly the wind responses of trees, buildings, and utility standards provide the surface waves this method analyzes.

Keywords: Microtremors · Catacombs · Underground tombs · Seismic assessment · Alexandria · ReMe test

1 Introduction

Microtremors (also called "microseisms") are seismic waves of relatively low energy having amplitudes typically in the range of 10–4 to 10–2 mm (Okada 2003). In general, microtremors are an assemblage of body and surface wave motions, although most of the wave energy is transported as surface waves (Toksöz and Lacoss 1968). To a first approximation, microtremors with a frequency greater than 1 Hz are produced by cultural sources (such as trains, road traffic, and machinery), while frequencies less than 1 Hz are the result of natural phenomena such as wave action at coastlines, wind, and atmospheric variations (Okada 2003).

Most seismic prospecting and seismological applications use the traditional "raypath" approach to estimate seismic velocity based upon the time taken for a distinct non-dispersive seismic "event" to propagate between two (or more) points of observation. However, microtremors are dispersive and form a continuous low amplitude wave "field" – an assemblage of body and surface waves that originates in space and time from a wide variety of sources, and propagates over a wide frequency band. While the amplitude and frequency content of microtremors can display significant variation in both space and time, they can be assumed stationary when considered over suitably short time intervals. As we have said before, microtremors are omnipresent low amplitude oscillations (1–10 μm) that arise predominantly from oceanic, atmospheric, and cultural disturbances. It may be considered to compose of any of seismic wave types. We have two main types of microtremors, local cultural noise coming from urban disturbances and long period microtremors originated from farther distances (e.g. oceanic disturbances).

As we have said before, microtremors are omnipresent low amplitude oscillations (1–10 μm) that arise predominantly from oceanic, atmospheric, and cultural disturbances. It may be considered to compose of any of seismic wave types. We have two main types of microtremors, local cultural noise coming from urban disturbances and long period microtremors originated from farther distances (e.g. oceanic disturbances).

Still, there is a big contradiction between the correct types of noise that should be used (especially for soil response). While some is dealing with the longer period Microtremors originated from farther distances (oceanographic disturbances) excluding Urban excited sites where high degree of local cultural noise exist (e.g. Field et al. 1990), others considered the traffic and cultural noise dose not affect microtremors response. Moreover, they considered the excited soils (sometimes by a helicopter) give even more convincing site responses. They found good agreement between soil responses excluded from excited soils compared with those derived from strong motion recorded data (Mucciarelli 1998). Mucciarelli (1998) made constrictions to microtremors generated by winds or coming out from asphalt (asphaltic waves) and considered them as false effects and should be removed.

The implicit assumption of early studies was that microtremors spectra are flat and broadband before they enter the region of interest. When microtremors enter preferable body it changes and resonate depending on the nature of the material, shape, and any other characteristics of this body. We think that there is no big difference between soil response and building response. The soil response is depending mainly on identifying the resonance of surface layering by dividing the microtremors response for surface layer by microtremors response of a nearby deeper bedrock called the "reference site" (outcrop of nearby consolidated sediments or basement rock) this is to maintain the fundamental resonance modes of vibrations for This beds. The reality of method is depending on the phenomenon that when both sites is having the same source of noise then by dividing them the effect of source of noise and path effect will vanish leaving the effect of the site. If this stands for the buildings too then by dividing the micro-tremors response of any floor (may be corridor, balcony or any other part of the structure) by the Microtremors response of the basement of the foundation, will exclude the effect of the source and path of the noise leaving the footprint of the floor alone (natural modes of vibrations and corresponding amplification).

Kanai (1957) first introduced the use of microtremors, or ambient seismic noise, to estimate the earthquake site response (soil amplification). After that lots of people followed this work but from the point of soil amplification of earthquake energy for different frequencies (e.g. Kanai and Tanaka 1961; Kanai 1962; Kagami et al. 1982, 1986; Rogers et al. 1984; Lermo et al. 1988; Celebi et al. 1987). We have used the same hypothesis to find the natural frequencies of vibration for buildings supposing that the same phenomenon will stand for buildings (Figs. 1a and 1b).

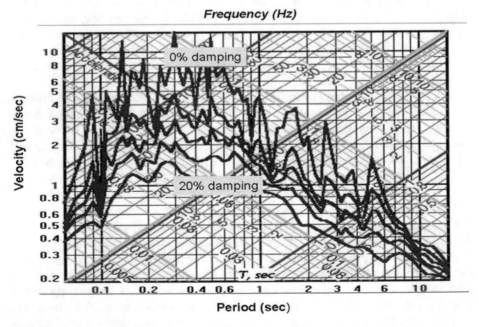

Fig. 1(a). Response spectra for different damping (0%, 2%, 5%, 10% and 20%) Determined from, Faiyum, Egypt, 12/10/1992 earthquake (Gamal 2001)

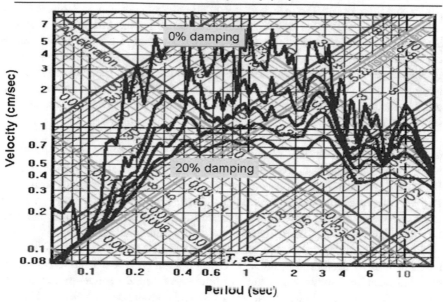

Fig. 1(b). Response spectra for different damping (0%, 2%, 5%, 10% and 20%) Determined from, Aqaba, Egypt, 22/11/1995 earthquake (Gamal 2001).

2 Background Theory and Method of Refraction Microtremor (REMI) Method

The need for the rapid and inexpensive assessment of Vs at large numbers of sites has led to the development of several geophysical testing methods that do not require drilling and are less laborious. The Spectral Analysis of Surface Waves (SASW) and microtremors array techniques both use surface-wave phase information to interpret shear-velocity or rigidity profiles. The refraction microtremors technique is based on two fundamental ideas. The first is that common seismic-refraction recording equipment, set out in a way almost identical to shallow P-wave refraction surveys, can effectively record surface waves at frequencies as low as 2 Hz. The second idea is that a simple, two-dimensional slowness-frequency (p-f) transform of a microtremors record can separate Rayleigh waves from other seismic arrivals, and allow recognition of true phase velocity against apparent velocities.

The SASW technique, sometimes referred to as "CXW" (Boore and Brown 1998), first introduced by Nazarian and Stokoe (1984), uses an active source of seismic energy, recorded repeatedly by a pair of 1 Hz seismometers at small (1 m) to large (500 m) distances (Nazarian and Desai 1993). The seismometers are vertical particle-velocity sensors, so shear-velocity profiles are analyzed on the basis of Rayleigh-wave phase velocities interpreted from the recordings. The phase velocities are derived purely from a comparison of amplitude and differential phase spectra computed from each seismometer pair for each source activation within an FFT oscilloscope (Gucunski and Woods 1991).

Since all interpretation is performed in the frequency domain, the SASW method assumes that the most energetic arrivals recorded are Rayleigh waves. However, when noise overwhelms the power of the artificial source, as is common in urban areas, or when body-wave phases are more energetic than the Rayleigh waves, SASW will not yield reliable results (Brown 1998; and Sutherland and Logan 1998). The velocities of Rayleigh waves cannot be separated from those of other wave types in the frequency domain. Boore and Brown (1998) found that SASW models consistently under-predicted shallow velocities, at the sites of six southern California borehole shear-velocity profiles.

In response to the shortcomings of SASW in the presence of noise, the Multi-channel Analysis of Surface Waves (MASW) technique (Park et al. 1999) was developed. The simultaneous recording of 12 or more receivers at short (1–2 m) to long (50–100 m) distances from an impulsive or vibratory source gives statistical redundancy to measurements of phase velocities. Multi-channel data displays in a time-variable frequency format also allow identification and rejection of non-fundamental-mode Rayleigh waves and other coherent noise from the analysis.

Miller et al. (2000) were able to obtain excellent MASW results in the noisy environment of an operating oil refinery. Using both large and stacked small sources, they could acquire records dominated by fundamental-mode Rayleigh waves. They also attempted 2-dimensional profiling for lateral anomalies in shear velocity by inverting many records along a profile. Such a profile represents much costly effort, similar to that needed for a high-resolution reflection survey, as a large source must be moved along and activated repeatedly at a large number of locations (Louie 2001).

The need for urban-friendly, reliable, low cost techniques, from an economical as well as from a safety point of view has led over the last decade to the development of passive (not requiring seismic sources) seismic techniques utilizing ambient vibration (microtremors) measurements, which are very easy to obtain even in the most urbanized areas.

Ambient vibration techniques are much cheaper than classical geophysical site investigations. In addition, the latter investigations are often almost impossible in urban environments, where, for instance, use of explosives is either forbidden or very strictly regulated, or the level of noise and artifacts prevents the use of refined techniques. That is why the use of the former techniques is rapidly spreading world-wide, especially again in urban areas.

In 2001, the so-called refraction microtremors technique or ReMi (Louie 2001) was introduced. This approach is also a passive seismic approach in which records of ambient vibrations by traffic, wind, is obtained and inverted in terms of Vs profiles to depths reaching 100 m. Being essentially a noise recording and analysis method, it works successfully inside buildings, over pavements, on roads and highways with exceptional ease. It has combined the shallow high resolution SASW and the field procedure of MASW. As such it can provide for rapid, cost-effective spatial coverage, and ease of conduct.

As opposed to the traditional techniques of estimating shallow shear velocities which are too costly to be employed for detailed studies, require large sources to be effective in noisy urban settings, or specialized independent recorders laid out in an extensive array, the refraction microtremors technique is very fast and inexpensive; it requires only standard refraction equipment; it requires no seismic energy source; and it will work best in a seismically noisy urban setting (The electrical power plant is a suitable environment for this test, because of the existence of continuous vibrating machines). Traffic and other vehicles, and possibly the wind responses of trees, buildings, and utility standards provide the surface waves this method analyses (Louie 2001).

There are several methods that can provide shear-wave velocity from microtremors (Aki 1957; Louie 2001; Okada 2003). It is impractical to use the circular or two-dimensional array (Okada 2003; Roberts and Asten 2004; Roberts and Asten 2005; Hayashi et al. 2004) in our case because the site is spatially restricted. Therefore, we decided to apply the REMI technique, which uses microtremors recorded with a linear array (Louie 2001). At these sites, abundant microtremors were generated by vehicles on the roads parallel or perpendicular to the sites, since el-shatbi and moustafa kamil areas are very close to the sea shoreline and Alkornish Street.

Louie (2001) proposed the refraction microtremor method that can provide shear-wave velocity to 100 metres depth, using conventional refraction equipment and a linear geophone array (Rucker et al. 2003; Pullammanappallil et al. 2003; Louie et al. 2002). In the inversion procedure, a Rayleigh-wave dispersion curve is picked in the wave field-transformed domain, and the subsurface shear-wave velocity profile can be determined by a processing procedure similar to that used for the multi-channel analysis of surface waves (MASW; Park et al. 1999).

ReMi processing involves three steps: Velocity Spectral Analysis, Rayleigh Phase Velocity Dispersion Picking and Shear-Wave Velocity Modeling.

2.1   Velocity Spectral Analysis

The basis of the velocity spectral analysis is the p-tau transformation, or "slantstack," described by Thorson and Claerbout (1985). This transformation takes a record section of multiple seismograms, with seismogram amplitudes relative to distance and time (x-t), and converts it to amplitudes relative to the ray parameter p (the inverse of apparent velocity) and an intercept time tau. It is familiar to array analysts as "beam forming," and has similar objectives to a two-dimensional Fourier-spectrum or "F-K" analysis as described by Horike (1985). Clayton and McMechan (1981) and Fuis et al. (1984) used the p-tau transformation as an initial step in P-wave refraction velocity analysis. The distinctive slope of dispersive waves is a real advantage of the p-f analysis. Other arrivals that appear in microtremors records, such as body waves and air waves, cannot have such a slope. The p-f spectral power image will show where such waves have significant energy. Even if most of the energy in a seismic record is a phase other than Rayleigh waves, the p-f analysis will separate that energy in the slowness-frequency plot away from the dispersion curves this technique interprets. By recording many channels, retaining complete vertical seismograms, and employing the p-f transform, this method can successfully analyze Rayleigh dispersion where SASW techniques cannot (Louie 2001).

2.2   Rayleigh Phase-Velocity Dispersion Picking

This analysis adds only a spectral power-ratio calculation to McMechan and Yedlin's 1981) technique, for spectral normalization of the noise records. The ability to pick and interpret dispersion curves directly from the p-f images of spectral ratio parallels the coherence checks in the SASW technique (Nazarian and Stokoe 1984) and the power criterion in the MASW technique (Park et al. 1999). Picking phase velocities at the frequencies where a slope or a peak in spectral ratio occurs clearly locates the dispersion curve. Picks are not made at frequencies without a definite peak in spectral ratio, often below 4 Hz and above 14 Hz where an identifiable dispersive surface wave does not appear. Often, the p-f image directly shows the average velocity to 30 m depth, from the phase velocity of a strong peak ratio appearing at 4 Hz, for soft sites, or nearer to 8 Hz, at harder sites.

Picking is done along a "lowest-velocity envelope" bounding the energy appearing in the p-f image. It is possible to pick this lowest-velocity envelope in a way that puts confidence limits on the phase velocities, as well as on the inverted velocity profile. Picking a surface-wave dispersion curve along an envelope of the lowest phase velocities having high spectral ratio at each frequency has a further desirable effect. Since higher mode Rayleigh waves have phase velocities above those of the fundamental mode, the refraction microtremor technique preferentially yields the fundamental-mode velocities. Higher modes may appear as separate dispersion trends on the p-f images, if they are nearly as energetic as the fundamental.

Spatial aliasing will contribute to artifacts in the slowness-frequency spectral-ratio images. The artifacts slope on the p-f images in a direction opposite to normal-mode dispersion. The p-tau transform is done in the space and time domain, however, so even the aliased frequencies preserve some information. The seismic waves are not continuously harmonic, but arrive in groups. Further, the refraction microtremor analysis has not just two seismograms, but 12 or more. So severe slowness wraparound does not occur until well above the spatial Nyquist frequency, about twice the Nyquist in most cases.

2.3   Shear-Velocity Modeling

The refraction microtremor method interactively forward shears wave velocity-models. The normal-mode dispersion data picked from the p-f images with a code adapted from Saito (1979, 1988) in 1992 by Yuehua Zeng. This code produces results identical to those of the forward modeling codes used by Iwata et al. (1998), and by Xia et al. (1999) within their inversion procedure. The modeling iterates on phase velocity at each period (or frequency), reports when a solution has not been found within the iteration parameters, and can model velocity reversals with depth.

3 Used Equipments

The used instrument is the "Start View" seismograph (Fig. 2, Geometrics Company, USA-www.geometrics.com) 24 bit A/D converter sampling at 31 1/4 microsecond interval, with a floating point number fed to a 32-bit, digital signal processing which performs multi-sample processing. The geophones used are 12 Hz single P-wave geophones (Singe-channel vertical-motion seismometers).

Amplitude or frequency-response calibration of geophones is not needed - as with refraction, ReMi uses only the phase information in the recorded wave field, the geophone cable is laid out on straight stretch of flat ground at the site and should be centered on the desired target. It is best to avoid known underground cavities 10 ft or more in diameter - pass beside but not over them. Geophones can be placed on thin pavements as long as they can be set so that there is good coupling with the ground. An easy urban layout is to run the array along the sidewalk, with the geophones planted into the parking strip or cracks in pavements. If the seismic cable must cross a street or driveways that cannot be blocked during the survey, put it between 2 × 4 s nailed to the pavement. For recording noise records, a deviation in the line of 5% of the total length will not affect the stated 15% velocity accuracy of the method. This accuracy applies to elevations as well - in fact the line can have a constant inclination that can safely be ignored, as long as geophone elevations do not deviate more than 5% from the incline. The geophone locations need to be surveyed in only if the array deviates more than 5% of the total length from a straight line or if the elevation changes more than 5% from a constant include.

Fig. 2. Used recording seismograph Start View seismograph (Geometrics company, USA).

4 Methodology

The data acquisition and analysis in the refraction microtremors technique follow 4 main steps:

(1) Acquiring 20 s of microtremors data along a linear or crooked array of low frequency vertical geophones (8–12 Hz).
(2) Applying slowness (p)-ray parameter transformation or the slant-stack of Thorson and Claerbout (1985) to the acquired data.
(3) Fourier transforming the slowness-ray parameter wavefield into the slowness-frequency (f) wavefield following the scheme of McMechan and Yedlin (1981). The slowness-frequency wavefield transformation is particularly effective in allowing accurate picking of Rayleigh-wave phase-velocity dispersion curves despite the presence of waves propagating across the linear array at high apparent velocities, higher-mode Rayleigh waves, body waves, air waves, and incoherent noise.
(4) Performing velocity spectral analysis to obtain the true phase velocity of the dispersed wave trains.

Louie (2001) showed that microtremors noise recordings made along lines of seismic refraction equipment could estimate shear velocity with 15% accuracy down to a 100 m depth, matching the average velocity profile yielded by a suspension logger run in a 100 m-deep borehole, and could suggest structures beyond the 100 m logged depth of the hole. Similar results were obtained over a wide range of hard and soft sites. The refraction microtremors technique was successfully applied, either stand-alone or in conjunction with other geophysical techniques, in the geotechnical characterization of soils and bedrock, in subsurface fracture detection, and in subsurface cavity delineation Rucker (2003). Scott et al. (2004) successfully applied the technique for NEHRP-UBC-soil classification in a basin in Nevada, obtaining Vs 30 at 55 sites in about a week.

ReMi profiles were conducted at six locations within the site as shown in Fig. 3 below. A StrataView 24-channel seismograph was used and, at each location, the geophones were laid out along a straight line with 9 m inter-geophone separation. This provided for a profile length of 207 m. At each location, 20 to 40 records of 20 s of ambient vibrations were recorded, providing for a total record approximately 7 to 13 min long.

The data collected was subjected to processing and inversion according to the scheme described previously to obtain Vs distribution. The Results are presented in the form of slowness (p)-frequency or phase velocity-frequency images, along with the inverted Vs profiles.

A) Catacomb of Kom El-shoqafa, ReMi-1.

B) Necropolis of Moustafa Kamil , ReMi-2.

C) El-Shatbi Necropolis , ReMi-3.

Fig. 3. Alexandria archaeological sites for Remi Tests Remi-1, Remi-2 and Remi-3, (a) Catacomb of Kom El-Shoqafa site, (b) Necropolis of Moustafa Kamil, (c) El-Shatbi necropolis.

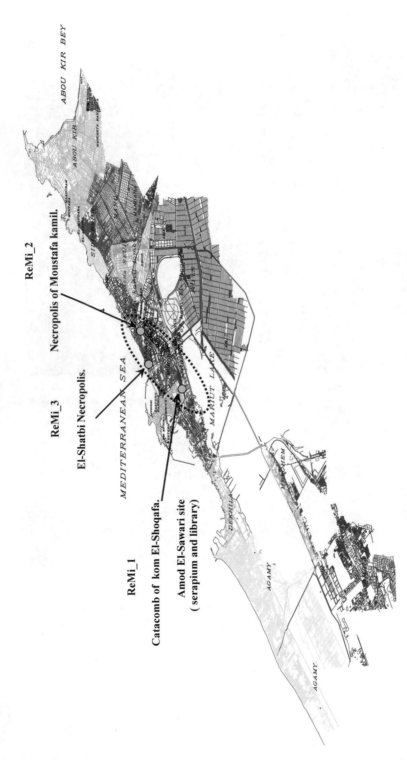

Fig. 4. Map of Alexandria, shows Location of the three archaeological sites under investigation.

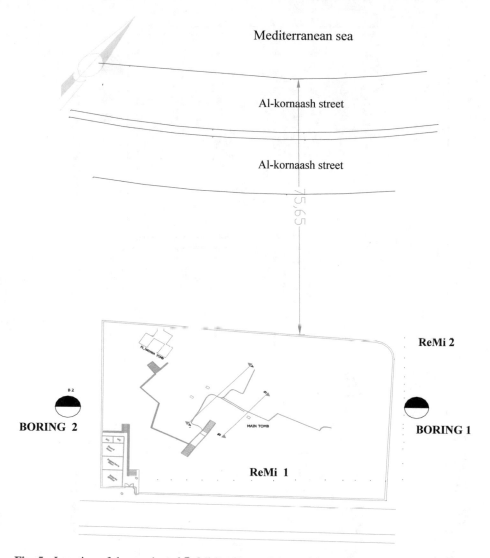

Fig. 5. Location of the conducted ReMi Profiles and the position of the geotechnical boreholes at El-shatbi necropolis.

14   S. Hemeda

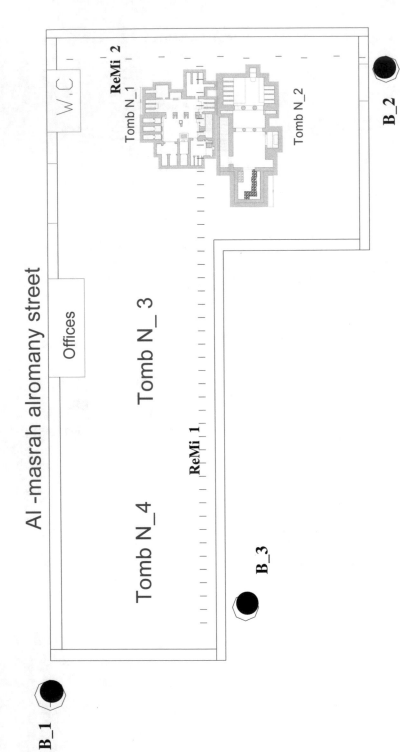

Fig. 6. Location of the conducted ReMi Profiles and the positions of the geotechnical boreholes at necrpolis of Moustafa Kamil.

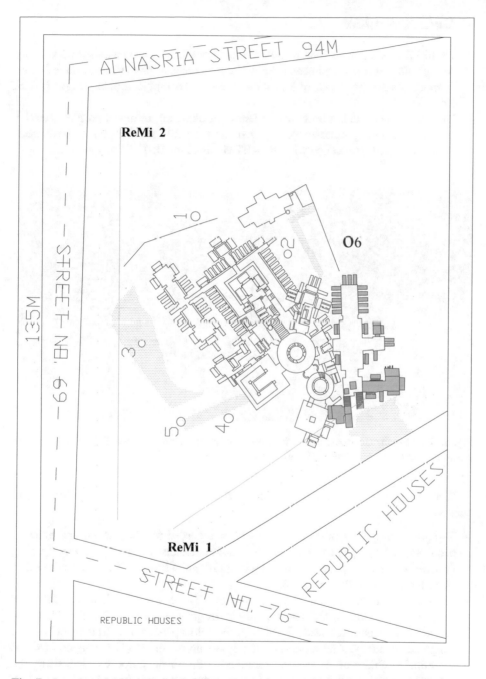

Fig. 7. Location of the conducted ReMi Profiles and the positions of the geotechnical boreholes at Catacomb of Kom El-Shoqafa site, Alexandria.

5   Data Acquisition

– Six ReMi profiles have been recorded to estimate the shear wave seismic velocity through the three monumental Sites in Alexandria city (Catacomb of kom El-shoqafa site, Necropolis of Moustafa kamil, El-shatbi Necropolis) (Figs. 3, 4, 5, 6 and 7).
– The geophone cable has been installed straight along all Remi test profiles (Remi-1 to Remi-6), with geophone spacing 9 m or total 207 m. Geophones have been installed in correspondence with the ASTM Standard D5777 (Figs. 4 and 8).

Fig. 8. Planting the Geophones along the Remi profiles with the good coupling soil (Site nature shows almost little or no loose soil/soft rock)

The geophones were installed almost vertical without compromising ReMi data quality.

– Twenty to forty files of 20 s each has been recorded for the background noise at each site, (20 s long each). Each record has 24 channels with 6 m geophone spacing (vertical P-wave geophones). The sampling interval was set to 2 ms (Each record of 20 s. length has 10000 samples).
– All filters were turned off before recording. The lowest possible low-cut filter frequency (less than 4 Hz) and a high-cut frequency equal to 250 Hz. This procedure is adequate to assure ReMi's 15% velocity accuracy. The records were triggered manually. And a passage of a good noise source (The compressors and vibrating machines of the station) were recorded.- for the purpose of increasing the amount of noise a truck was derived up and down the geophone line.

6   Interpretation Softwares

The used software is the SeisOpt ReMi software which uses refraction microtremors recordings from standard refraction equipment to estimate the average shear wave velocities. With a 5% to 15% accuracy. The Software was used for data handling and interpretation (2006 version).

7   Results and Discussions

Figure 9 through Fig. 11 shows the Rayleigh-wave phase-velocity dispersion picks, p-f Image using SeisOpt ReMi with Dispersion Modeling Picks while Table 1 through Table 3 and Fig. 12 Through Fig. 14 Show the excluded shear-wave seismic velocity models for all Remi test profiles (Remi-1 to Remi-3) obtained by modeling the phase-velocity picks shown in Fig. 9 through Fig. 11.

The obtained shear wave seismic velocities for all Remi tests Remi-1 to Remi-6 show a relatively high range of shear wave velocities ranging between 260 m/s to 1420 m/s for the Remi_1 catacomb of (kom El-shoqafa site), and for Remi_2 (moustafa kamil tombs site) is between 450 m/s to 1600 m/s, and for the third site(El-shatbi tombs site) is between 2401 m/s to 1500 m/s, for the three p-f images we note how the velocity inversions are apparent, the square boxes represent picks that are modeled to obtain the shear wave profiles shown in Fig. 12 through 14. This is evidenced from the surface soil.

The obtained shear wave seismic velocities for Remi tests show a relatively high range of shear wave velocities ranging between 260 m/s to 1420 m/s. However it is clear that the ground conditions were Catacombs and other tombs are excavated cannot be classified as real rock at least close to the surface.

Note that disparity for certain depths is due to buried ancient structural members who found at the three sites.

– **ReMi_1 (Catacomb of kom El-shoqafa site)**

The p-f analysis of this 30-s noise record is Fig. 9. From 2–10 Hz, the pf image shows a clear energy cutoff at a minimum velocity envelope. The energy of obliquely propagating waves is broadly distributed across high apparent velocities above this envelope. Arrivals at many different apparent velocities form a broad ramp in spectral ratio, but the cutoff of high spectral-ratio values against the true phase-velocity envelope is clear from 2–10 Hz. At frequencies below 2 Hz, and in the area of F-K aliasing, this envelope is not as clear. There are a few spectral ratio peaks in these areas still aligned with the dispersion envelope.

The dispersion picks follow the lowest-velocity envelope at the base of the high spectral ratios in the image (Fig. 9). The area of pick confidence is between the lowest velocity where spectral ratios rise above those of uncorrelated noise and the higher velocity at the top of the ratio peak). The "best pick" was made within this range where the ratio slope is steepest.

Velocity modeling - Fig. 9 shows just the dispersion curve with increasing velocity uncertainty at larger periods, interactively modeled with the velocity profile of Fig. 12

(bold line). The 2-m-thick shallowest layers is with 265 m/s shear velocity, from 2 m to 4 m-depth has a velocity of 320 m/s. Below 4 m until 6.5 m, the shear velocity log has high variability is increase to 1150 m/s, and reverse to be 410 m/s from 6 m to 9 m depth, but increase to 720 m/s in depth from 9 to 13 m, and from 13 m to 30 m depth increased highly to 1420 m/s.

As shown in Fig. 12, the subsurface appears to be composed of 4 predominate strata described as follows: The shear-wave profile describes a thin low velocity surficial soil to a depth of about 2 m. A moderate velocity stratum extending from 2 m to a depth of about 4 m where the velocity profile increases underlies these surficial soils. The stratum in the 4 m to 16 m zone contain the underground structures of the catacomb and contain at least one thin, high velocity layer from 4 m to 6 m depth. The primary velocity reversal is likely due to the presence of either a very low strength granular soil or more likely a silty, clayey material. The velocity gradually increases to speeds indicative of a transition into soft bedrock at 10 m to 13 m, from 13 m to 40 m or more with high velocity about 1430 m/s. Hard rock is likely a bit deeper than 30 feet. The 24 receiver array provided confident data no deeper than about 40 m.

The assumption of an average profile is consistent with the limited area and uncertainties in engineering properties of individual strata, as follows.

1- Filling of dirty grey crushed stones, crushed bricks filling of dirty yellow crushed cal. cemented sand, some med/coarse cal. Sand, filling of dirty yellow pieces of crushed cal. cemented sand. From 0.00 to 4.00 m depth
2- Filling of dirty yellow med/coarse cal. Sand, trace pieces of cal .cemented sand, trace fine crushed bricks. Filling of dirty grey crushed cal. stones crushed brick, little sand. From 4.00 to 10.00 m depth
3- Filling of dirty, grey pieces of crushed cal. Stone crushed bricks. From 10.00 to 12.00 m depth
4- Yellow med/coarse cal. sand trace pebbles of cal. cemented sand, trace very fine crushed shells. Yellow med/coarse cal. sand, crushed cal. cemented sand, trace fine crushed shells. From 12.00 to 25 m depth.

– **ReMi_2 (Necropolis of Moustafa kamil)**

P-f-image obtained using SeisOpt ReMi, shows how the velocity inversions are apparent in the image; the square boxes represent picks that are modeled to obtain the shear wave profile shown in Fig. 13. The dispersion picks follow the lowest-velocity envelope at the base of the high spectral ratios in the image (Fig. 10). The area of pick confidence is between the lowest velocity where spectral ratios rise above those of uncorrelated noise and the higher velocity at the top of the ratio peak). The "best pick"' was made within this range where the ratio slope is steepest.

Velocity modeling - Fig. 10 shows just the dispersion curve with increasing velocity uncertainty at larger periods, interactively modeled with the velocity profile of Fig. 13 (bold line).

From the Vs model we found that, The 2-m-thick shallowest layers is with 460 m/s shear wave velocity, and the stratum from 2 m to 9 m-depth has high steeply variability is increases from 460 to 780 to 1110 to 1600 m/s shear wave velocity only in 2 m zone and continues at this value 1600 m/s until the end of this stratum 9 m depth. Beneath these layers the shear velocity log has high variability is reverse to 300 m/s from 9 m to 13 m depth, but increases again to 1250 m/s in layer from 13 m to 16 m depth, and from 16 m until the end of ReMi test 30 m and more, reverse again to 640 m/s. This disparity for certain depths is due to buried ancient structural members who found at the site.

As shown in Fig. 13, the subsurface appears to be composed of 3 predominate strata described as follows: The shear-wave profile describes a thin low velocity surficial soil to a depth of about 2 m. A high velocity rock stratum extending from 2 m to a depth about 9 m where the velocity profile increases underlies these surficial soils (this stratum includes the underground structures of the tombs 1 and 2). The stratum from 2 m to 9 m depth zone could contain at least three thin, high velocity layers. The primary velocity reversal is likely due to buried ancient structural members who found at the site, and due to the presence of either a very low strength granular soil or more likely a silty, clayey material. after 9 m depth the velocity extremely decreases to speeds indicative of a transition into soft bedrock at 10 m to 14 m to be 300 m/s, and increases again to 1250 m/s in the layer between 14 to 17 m depth, but reverse again to be 640 m/s from 17 m depth until the end of layer which more than 40 m the end of profile.

The assumption of an average profile is consistent with the limited area and uncertainties in engineering properties of individual strata, as follows.

1- filling of dark yellow poorly graded sand with hard dark brown calcareous silt with yellow detrital sandy oolitic limestone and little pieces of building ruins. From 0.00 m to 2.00 m depth
2- yellow poorly graded sand with hard dark calcareous silt with yellow detrital sandy oolitic limestone. From 2.00 m to 9.00 m depth
3- yellow detrital sandy oolitic limestone with little calcareous pebbles and monumental pottery pieces. From 9.00 m to 10.00 m depth
4- poorly yellow detrital sandy oolitic limestone with little iron oxides. From 10.00 m to 14.00 m depth.

– ReMi_3 (El-shatbi necropolis)

P-f-image obtained using SeisOpt ReMi, shows how the velocity inversions are apparent in the image; the square boxes represent picks that are modeled to obtain the shear wave profile shown in Fig. 14. The dispersion picks follow the lowest-velocity envelope at the base of the high spectral ratios in the image (Fig. 11). The area of pick confidence is between the lowest velocity where spectral ratios rise above those of uncorrelated noise and the higher velocity at the top of the ratio peak). The "best pick" was made within this range where the ratio slope is steepest.

Velocity modeling - Fig. 11 shows just the dispersion curve with increasing velocity uncertainty at larger periods, interactively modeled with the velocity profile of Fig. 14 (bold line). The 2-m-thick shallowest layers is with 230 m/s shear wave velocity, the layer from 2 m to 4 m-depth has high variability is increase to 1020 m/s shear wave velocity. Below 4 m until 6.5 m, the shear velocity log has high variability is reverse to 350 m/s, but increased again to 1520 m/s in layer from 6 to 9 m depth, and from 9 m to 12 m depth increased little to be 1540 m/s, at depth 12 m, it reverse again to be 1220 m/s until the end of ReMi test 30 m and more.

This disparity for certain depths is due to buried ancient structural members who found at the site.

As shown in Fig. 14, the subsurface appears to be composed of 4 predominate strata described as follows: The shear-wave profile describes a thin low velocity surficial soil to a depth of about 2 m. A high velocity stratum extending from 2 m to a depth of about 5 m, the velocity profile reverses (decreases) underlies these surficial soils. The stratum from surface to 6 m depth zone could contain at least one thin, high velocity layer. The primary velocity reversal is likely due to the presence of either a very low strength granular soil or more likely a silty, clayey material. The velocity gradually increases to speeds indicative of a transition into soft bedrock at 6 m to 13 m to be 1520 m/s, and reverse again to 1230 m/s in the fourth stratum which extends from 13 m to 30 m or more with high velocity about 1230 m/s. Hard rock is likely a bit deeper than 30 feet. The 24 receiver array provided confident data no deeper than about 40 m. The depth between 2 m to 13 m include the underground structures of El-shatbi tombs.

The assumption of an average profile is consistent with the limited area and uncertainties in engineering properties of individual strata, as follows.

1- Filling of dense dark yellow micaceous gravelly silty poorly graded sand and brown calcareous silt with homra and basalt pieces. From 0.00 m to 1.00 m depth.
2- Sand with hard dark brown calcareous silt with little calcareous pebbles and homra pieces and building ruins pieces. From 1.00 m to 5.50 m depth.
3- Yellow oolitic calcareous sand with yellow calcareous silt with little hard dark brown connected laminated sandy silt. From 5.50 m to 7.00 m depth.
4- Yellow detrital sandy fossiliferrous limestone with little iron oxides with little hard dark yellow calcareous silt. From 7.00 m to 12.00 m depth.
5- Yellow detrital sandy fossiliferrous limestone with little iron oxides. From 12.00 m to 15.00 m depth, the end of boreholes.

In general, shortening the receiver spacing will improve resolution at the sacrifice of depth of investigation, since the total array length is shortened. The natural frequency of the receiver can also influence resolution, as low frequency receivers will respond better to low frequency waves that sample deeper strata and are a benefit when depth of investigation is important. A higher frequency receiver will help to resolve shallow structure since high frequency wavelets that it records sample only the more local shallow structure. It is also likely the frequency of the geophones might determine the thickness of the layers that are resolved by ReMi. It is possible to perform a ReMi survey using a variety of receiver types then combine the data to analyze a larger range of frequencies to further improve resolution (Tables 2, 4, 5, 6 and Figs. 15, 16).

Table 1. The excluded Shear wave velocity depth model for Remi-1 test.

Depth, m	Vs, m/s
0	267.025
−2.331	267.025
−2.331	310.329
−4.429	310.329
−4.429	1149.799
−6.061	1149.799
−6.061	416.182
−9.324	416.182
−9.324	714.496
−12.821	714.496
−12.821	1425.339
−40.559	1425.339
−40.559	1093.36

Table 2. The excluded Shear wave velocity depth model for Remi-2 test.

Depth, m	Vs, m/s
0	456.504
−2.671	456.504
−2.671	781.707
−3.647	781.707
−3.647	1120.461
−4.783	1120.461
−4.783	1601.491
−9.255	1601.491
−9.255	287.127
−13.287	287.127
−13.287	1249.187
−16.272	1249.187
−16.272	640.386
−30	615

22   S. Hemeda

Table 3. The excluded Shear wave velocity depth model for Remi-3 test.

Depth, m	Vs, m/s
0	226.152
–2.098	226.152
–2.098	1018.835
–4.196	1018.835
–4.196	354.878
–5.828	354.878
–5.828	1513.415
–9.091	1513.415
–9.091	1533.74
–12.588	1533.74
–12.588	1235.637
–40.326	1235.637
–40.326	1012.06

Table 4. The excluded Shear wave velocity depth model and young's, shear modulus for Remi-1 test.

Depth, m	Vs, m/s	Young's modulus (E) KN/m^2	Shear modulus (G) KN/m^2
0	267.025	301181.129	192755.9
–2.331	267.025	301181.129	192755.9
–2.331	310.329	406788.4687	260344.6
–4.429	310.329	406788.4687	260344.6
–4.429	1149.799	5584287.415	3573944
–6.061	1149.799	5584287.415	3573944
–6.061	416.182	731628.2989	468242.1
–9.324	416.182	731628.2989	468242.1
–9.324	714.496	2156371.152	1380078
–12.821	714.496	2156371.152	1380078
–12.821	1425.339	8581441.503	5492123
–40.559	1425.339	8581441.503	5492123
–40.559	1093.36	5049522.042	3231694

Table 5. The excluded Shear wave velocity depth model and young's, shear modulus for Remi-2 test.

Depth, m	Vs, m/s	Young's modulus (E) KN/m^2	Shear modulus (G) KN/m^2
0	456.504	853589.6147	546297.4
–2.671	456.504	853589.6147	546297.4
–2.671	781.707	2502925.655	1601872
–3.647	781.707	2502925.655	1601872
–3.647	1120.461	5142252.964	3291042
–4.783	1120.461	5142252.964	3291042
–4.783	1601.491	10505311.94	6723400
–9.255	1601.491	10505311.94	6723400
–9.255	287.127	337682.0803	216116.5
–13.287	287.127	337682.0803	216116.5
–13.287	1249.187	6391677.587	4090674
–16.272	1249.187	6391677.587	4090674
–16.272	640.386	1679745.962	1075037
–30	615	1549209.6	991494.1

Table 6. The excluded Shear wave velocity depth model and young's, shear modulus for Remi-3 test.

Depth, m	Vs, m/s	Young's modulus (E) KN/m^2	Shear modulus (G) KN/m^2
0	226.152	225548.2465	142095.4
–2.098	226.152	225548.2465	142095.4
–2.098	1018.835	4577689.179	2883944
–4.196	1018.835	4577689.179	2883944
–4.196	354.878	555388.3214	349894.6
–5.828	354.878	555388.3214	349894.6
–5.828	1513.415	10100774.08	6363488
–9.091	1513.415	10100774.08	6363488
–9.091	1533.74	10373900.49	6535557
–12.588	1533.74	10373900.49	6535557
–12.588	1235.637	6733182.689	4241905
–40.326	1235.637	6733182.689	4241905
–40.326	1012.06	4517010.606	2845717

Site -1 : Supportive Illustration

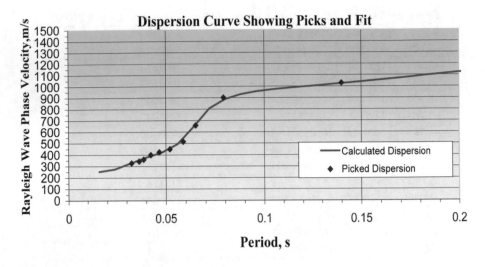

p-f Image with Dispersion Modeling Picks

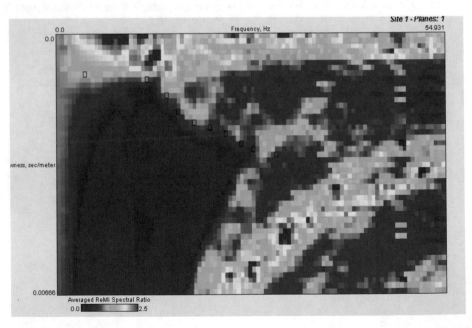

Fig. 9. The dispersive curve, P-f image and corresponding picks for Remi-1 test, Catacomb of Kom ElSshoqafa site.

Site - 2 : Supportive Illustration

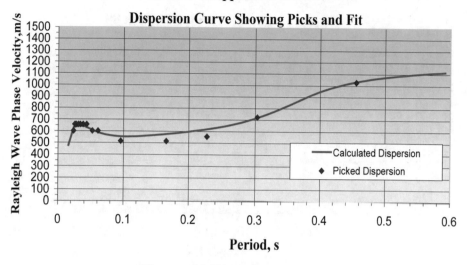

Dispersion Curve Showing Picks and Fit

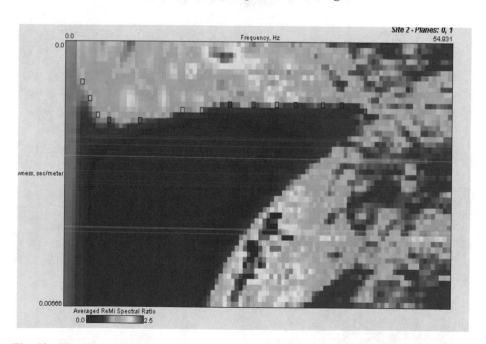

p-f Image with Dispersion Modeling Picks

Fig. 10. The dispersive curve, P-f image and corresponding picks for Remi-2 test, Necropolis of Moustafa Kamil.

Site - 3 : Supportive Illustration

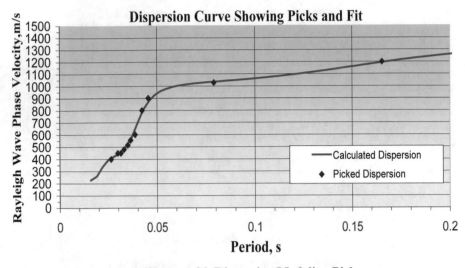

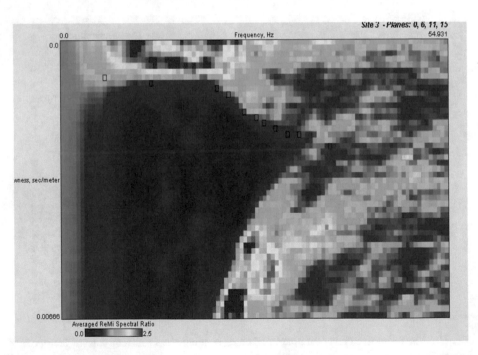

Fig. 11. The dispersive curve, P-f image and corresponding picks for Remi-3 test, El-Shatbi Necropolis.

Site-1: Vs Model

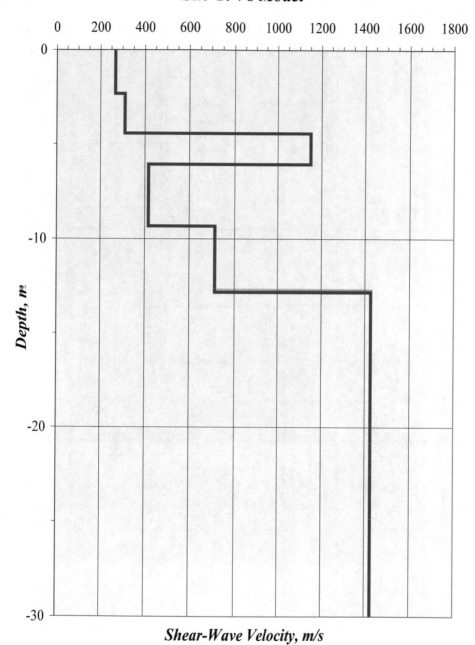

Fig. 12. The excluded Shear wave velocity model for Remi-1 test, Catacombs of Kom El-shoqafa (Fig. 7)

Site-2: Vs Model

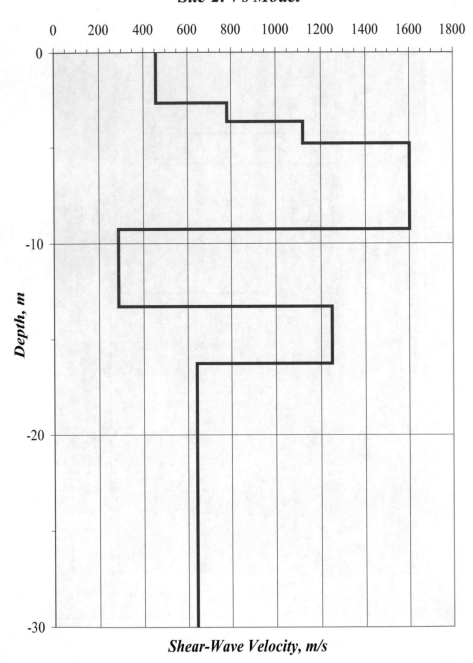

Fig. 13. The excluded Shear wave velocity model for Remi-2 test, Necropolis of Moustafa Kamil (Fig. 6)

Site-3: Vs Model

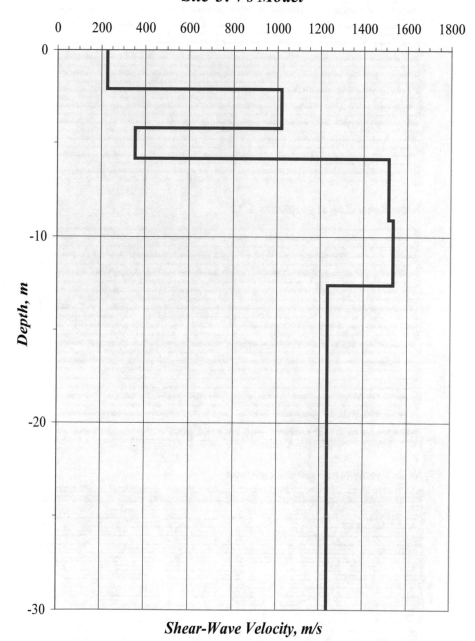

Fig. 14. The excluded Shear wave velocity model for Remi-3 test, El-Shatbi Necropolis. (Fig. 5)

30   S. Hemeda

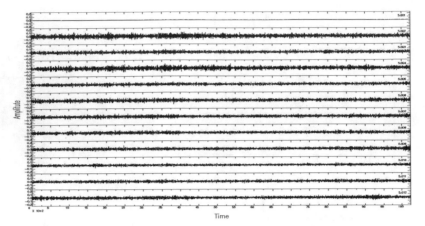

Noise Recording at geophones 1-12

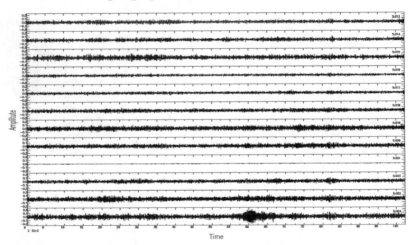

Noise Recording at geophones 13-24

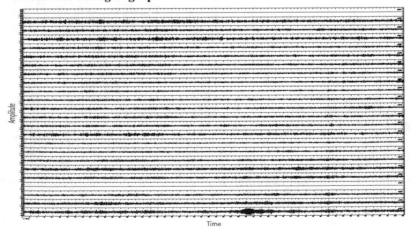

Fig. 15. Microtremors recording at moustafa kamil tombs area at geophones 1–24, with min amplitude –0.25 and max amplitude 0.25 for every geophone.

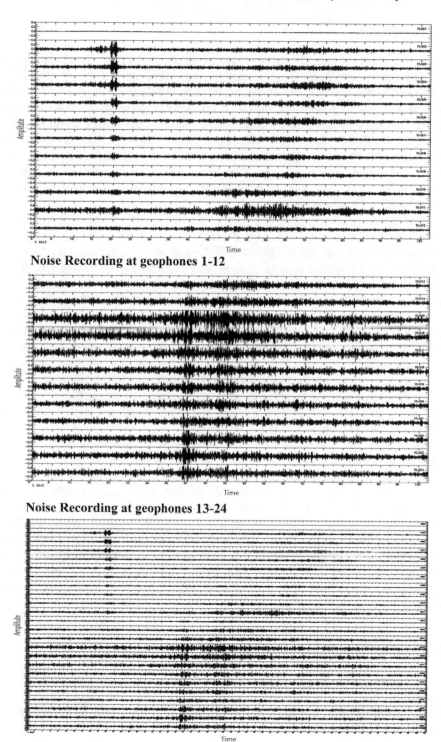

Noise Recording at geophones 1-12

Noise Recording at geophones 13-24

Fig. 16. Microtremors recording at El-shatbi tombs area at geophones 1–24, with min amplitude −0.4 and max amplitude 0.4 for every geophone.

8 Conclusion

The case studies tests have shown that common seismic refraction equipment can yield accurate surface-wave dispersion information from microtremor noise. Configurations of 12 to 48 single vertical, 4.5–14 Hz exploration geophones can give surface-wave phase velocities at frequencies as low as 1.5 Hz, and as high as 35 Hz. It is to be noted that the frequency range is a function of the frequency of the geophones used and the subsurface conditions. Presence of hard rock allows deeper penetration while thick alluvial fill will attenuate frequencies. One can see that dispersion picks can be easily made at frequencies less than 1 Hz, allowing velocities to be resolved to depths of 100 m. In this case picks can be made all the way out to 45 Hz allowing detailed resolution of shallow velocities.

Thus the depth of constrain will vary depending on the frequencies that can be resolved, but for most cases velocities down to 30 m can be resolved. If conditions allow a depth of 100 m can be achieved. If the goal is to get an average shear-wave profile, the heavy triggered sources of seismic waves used by the SASW and MASW techniques to overcome noise are not needed, saving considerable survey effort. This microtremor technique may be most fruitful, in fact, where noise is most severe. Proof of this technique suggests that rapid and very inexpensive shear-velocity evaluations are now possible at the most heavily urbanized sites, and at sites within busy transportation corridors. The ReMi method can also be fine-tuned to offer detailed look at the subsurface structure. For example, shorter geophone spacing and higher-frequency phones can resolve shallow structure (<30 m) in great detail (4 to 5 layers). Lateral changes along the profile can also be investigated by selecting fewer traces (from the 12 or 24 channels recorded) to analyze. A 1-D profile can be derived for each section and then put together to get a 2-D image.

The ReMi method offers significant advantages. In contrast to borehole measurements ReMi tests a much larger volume of the subsurface. The results represent the average shear wave velocity over distances as far as 200 m. Because ReMi is non-invasive and non-destructive, and uses only ambient noise as a seismic source, no permits are required for its use. ReMi seismic lines can be deployed within road medians, at active construction sites, or along highways, without having to disturb work or traffic flow. Unlike other seismic methods for determining shear wave velocity, ReMi will use these ongoing activities as seismic sources. There is no need to close a street or shut down work for the purpose of data acquisition and a ReMi survey usually takes less than two hours, from setup through breakdown. These advantages sum to substantial savings in time and cost.

References

AIJ- The Architectural Institute of Japan: Earthquake Motion and Ground Conditions. AIJ, 596 p. (1993)

Aki, K.: Space and time spectra of stationary stochastic waves, with special reference to microtremors. Bull. Earthq. Res. Inst. **35**, 415–456 (1957)

Boore, D.M., Brown, L.T.: Comparing shear-wave velocity profiles from inversion of surface-wave phase velocities with downhole measurements; systematic differences between the CXW method and downhole measurements at six USC strong-motion sites. Seismol. Res. Lett. **69**, 222–229 (1998)

Borcherdt, R.D., Glassmoyer, G.: On the characteristics of local geology and their influence on ground motions generated by the Loma Prieta earthquake in the San Francisco Bay region, California. Bull. Seimol. Soc. Am. **82**, 603–641 (1992)

Brown, L.T.: Comparison of Vs profiles from SASW and borehole measurements at strong motion sites in southern California. M.Sc. Engineering Thesis, University Texas at Austin, 349 p. (1998)

Capon, J.: High-resolution frequency-wave number spectrum analysis. Proc. IEEE **57**, 1408–1418 (1969)

Cho, I., Nakanishi, I., Ling, S., Okada, H.: Application of forking genetic algorithm fGA to an exploration method using microtremors. BUTSURI-TANSA **52**(3), 227–246 (1999)

Clayton, R.W., McMechan, G.A.: Inversion of refraction data by wave field continuation. Geophysics **46**, 860–868 (1981)

Gamal, M.A.: Seismic hazard analysis of Egypt and seismic microzonation of the Greater Cairo based on empirical and theoretical models. Ph.D. Faculty of Science, Geophysics Department, Cairo University, Egypt (2001)

Gucunski, N., Woods, R.D.: Instrumentation for SASW testing. In: Geotechnical Special Publication No. 29: Recent Advances in Instrumentation, Data Acquisition, and Testing in Soil Dynamics, NY, pp. 1–16 (1991). Am. Soc. of Civil Engineers

Horike, M.: Inversion of phase velocity of long period microtremors to the S-wave-velocity structure down to the basement in urbanized areas. J. Phys. Earth. **33**, 59–96 (1985)

Horita, J., Kita, K., Sasaki, M., Sakata, Y., Horiuchi, Y., Okada, H.: Application of surface wave method to the determination of subsurface structure - appraisal of field tests, Technical Reports of Hokkaido Branch, Japanese Geotechnical Society, No. 39, pp. 59–60 (1999)

Imai, T.: P- and S-wave velocities of the ground in Japan. In: Proceedings of 9th ISCMFE, Tokyo, vol. 2, pp. 257–260 (1981)

Iwata, T., Kawase, H., Satoh, T., Kakehi, Y., Irikura, K., Louie, J.N., Abbott, R.E., Anderson, J. G.: Array microtremor measurements at Reno, Nevada, USA (abstract). EOS Trans. Am. Geophys. Union **79**(45), F578 (1998)

Japan Road Association: Specifications for Highway Bridges. Part V, Earthquake Resistant Design (1990). (in Japanese)

Kramer, S.L.: Geotechnical Earthquake Engineering, 653 p. Prentice Hall, Upper Saddle River (1996)

Lacoss, R.T., Kelly, E.J., Toksoz, M.N.: Estimation of seismic noise structure using arrays. Geophysics **34**, 21–38 (1969)

Ling, S.: Studies of estimating the phase velocity of surface waves in microtremors, Faculty of Science, Hokkaido University (1994)

Ling, S., Shiono, T., Saito, F.: The evaluation improvement effect of soft subsoil for the compact vacuum consolidation method by using high precision surface wave prospecting method. In: The Sino-Japanese Symposium on Geotechnical Engineering, Beijing, China, pp. 142–147, 29–30 October 2003

Liu, Y., Wang, Z., Tanaka, Y., Zhang, Z.: Development and field example of multi channel surface wave data acquisition and processing system (SWS-1). In: The 94th SEGJ Conference, pp. 207–210 (1996)

Liu, Y., Ling, S., Okada, H.: Estimation of a subsurface structure by using a shallow seismic engineering exploration system with multiple functions (SWS). In: The 96th SEGJ Conference, pp. 11–14 (1997)

Ling, S., Horiti, J., Noguchi, S.H.: Estimation of shallow S-wave velocity structure by using high precision surface wave prospecting and microtremor survey method. In: 13th World Conference on Earthquake Engineering, Vancouver, B.C., Canada, 1–6 August 2004 paper No. 1445 (2004)

Louie, J.N.: Faster, better: shear-wave velocity to 100 m depth from refraction microtremors arrays. BSSA **91**, 347–364 (2001)

McMechan, G.A., Yedlin, M.J.: Analysis of dispersive waves by wave field transformation. Geophysics **46**, 869–874 (1981)

Miller, R.D., Park, C.B., Ivanov, J.M., Xia, J., Laflen, D.R., Gratton, C.: MASW to investigate anomalous near-surface materials at the Indian Refinery in Lawrenceville, Illinois. Kansas Geol. Surv. OFR **4**, 48 (2000)

Nazarian, S., Stokoe II, K.H.: In situ shear wave velocities from spectral analysis of surface waves. In: Proceedings of the World Conference on Earthquake Engineering, San Francisco, California, vol. 8, 21–28 July 1984

Nazarian, S., Desai, M.R.: Automated surface wave method: field testing. J. Geotech. Eng. **119**, 1094–1111 (1993)

Park, C.B., Miller, R.D., Xia, J.: Multi-channel analysis of surface waves. Geophysics **64**, 800–808 (1999)

Roberts, C., Asten, W.: Resolving a velocity inversion at the geotechnical scale using the microtremor (passive seismic) survey method. Explor. Geophys. **35**, 14–18 (2004). Butsuri-tansa (vol. 57, no. 1) Mulli-Tamsa (vol. 7, no. 1) (2004)

Rucker, M.L.: Applying the refraction microtremors (ReMi) shear wave technique to geotechnical characterization. In: The 3rd International Conference on the Application of Geophysical Methodologies to Transportation Facilities and Infrastructure, Orlando, FL, 8–12 December 2003

Scott, J.B., Clark, M., Rennie, T., Pancha, A., Park, H., Louie, J.N.: A shallow shear-wave velocity transect across the Reno, Nevada Area Basin. BSSA **94**, 650–667 (2004)

Sutherland, A.J., Logan, T.C.: SASW measurement for the calculation of site amplification. Earthquake Commission Research Project 97/276: Unpub. Central Laboratories Report 98-522422, Lower Hutt, New Zealand, 22 p. (1998)

Thorson, J.R., Claerbout, J.F.: Velocity-stack and slant-stack stochastic inversion. Geophysics **50**, 2727–2741 (1985)

Xia, J., Miller, R.D., Park, C.B.: Estimation of near-surface shear-wave velocity by inversion of Rayleigh wave. Geophysics **64**, 691–700 (1999)

Monitoring of Marine Sands Before and After Vibroflottation Treatment

Khelalfa Houssam[1,2,3](✉) (iD)

[1] School of Civil Engineering and Surveying, Faculty of Technology,
University of Portsmouth, Portsmouth, UK
khelalfahoussam@gmail.com
[2] Civil Engineering and Environmental Laboratory (LGCE), University of Jijel,
Cité 800 lgmts C5-N°03, Taher, Jijel, Algeria
[3] Department of Civil, Geotechnical & Coastal Engineering of K.E.C
Laboratory, Cité 800 lgmts C5-N°03, Taher, Jijel, Algeria

Abstract. One of the major problems related to civil engineering structures is that of ground movements with amplitude ranges from a few millimeters to a few meters. Soil treatment by vibroflotation is a recent technique for improving soil with poor geo-mechanical properties. In addition, this treatment minimizes the risk of liquefaction and settlement. This Article is interested to establish diagnostic and monitoring of vibroflottation works, basing on the results of SPT tests, laboratory tests and quality control as well as the details of bathymetric and the properties of seabed soils before and after treatment. On the other hand, the work consists in making a in-situ monitoring using topographic hardware carried out on the treated soil, in order to verify its influence on the stability of the vertical breakwaters. The tests results will be compared with in-situ measurements.

Keywords: Coastal sands · Vibroflottation · SPT · Liquefaction · Monitoring · Settlement

1 Introduction

Soil is usually a heterogeneous material with very variable characteristics. The main problems related to soils in general are manifested by a low bearing capacity, large deformations (absolute or differential settlements) under static or dynamic loads (earthquake, liquefaction), especially for loose and saturated sandy soils [1]. Thus, more and more buildings and infrastructures are built on soils of poor quality such as soft soils in coastal areas or on marshy sediment deposition areas [2]. The problem of building structures in low bearing lground is a current problem. This leads to a growing importance of methods and soil improvement technologies [3].

Vibroflotation is one of the most competitive soil improvement methods because of its speed of execution and its competitive price compared to other existing methods [4, 5]. Treatment with this method generally achieves the following goals:

© Springer Nature Switzerland AG 2020
H. Shehata et al. (Eds.): GeoMEast 2019, SUCI, pp. 35–56, 2020.
https://doi.org/10.1007/978-3-030-34184-8_2

- increasing the bearing capacity,
- the reduction of settlement,
- acceleration of consolidation,
- eliminating the risk of liquefaction,
- no adverse effects have been reported on the environment.

It is thanks to these practical advantages that the vibroflotation has become of intensive use on the international scale for economic reasons (cost, time of execution) [6] compared to other solutions such as the deep foundations or other [7, 8].

2 Improvement of Granular Soils

The high permeability of the granular soils leads to an increase in the interstitial pressures in these soils only in the case of liquefaction, usually caused by seismic stresses. The problems of granular soils are essentially problems of settlement amplitude, as well as problems of resistance to liquefaction, these problems arise mainly in the case of loose sands [9]. Compacting involves applying enough energy to the ground to reduce the void ratio and thus increase compactness (densify the soil in place) [3, 10], the objective is to increase shear strength and decrease permeability to stabilize the soil to increase bearing capacity and reduce settlement to withstand high loads [11, 12].

2.1 Vibroflotation Technique

Vibroflotation is a technique for in situ densification of thick layers of loose granular soil deposits (Figs. 1 and 2). It consists of generating, with the aid of a vibro-depth vibrator (vibroflot), horizontal vibrations in the granular soils in order to shear them and cause a localized liquefaction and an immediate settlement [13].

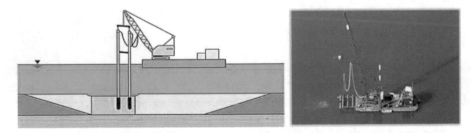

Fig. 1. Vibroflotation device on the barge [3, 10].

The vibroflotation method uses compression waves to compact the soil, rearranging the grain distribution pattern while applying cyclic vibration. As a result, soil compaction and pore volume reduction are achieved [14].

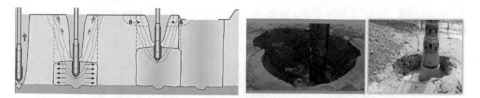

Fig. 2. The steps of the vibroflotation operation and the sinking of the stem (the vibrator is lowered on the seabed to the point of compaction) [10, 11].

During the vibroflotation process, vibrations in the soil allow the soil particles to rearrange under the forces of gravity in the densest possible state. This increases the relative density of the treated soil body, resulting in an overall reduction in volume [15].

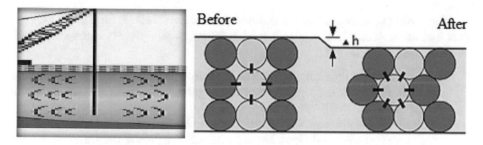

Fig. 3. Principle of vibrations and Rearrangement of sand particles [3].

The granular medium begins to dilate and gradually compact. The consolidation begins just after the pore pressure level reaches its maximum value in the sediment, the grains then redeposit gradually from the deepest liquefied zone. At the end of liquefaction, the surface of the sediment is rigid again and no longer has any ripples. The granular medium will eventually be compacted compared to the initial state [16, 17] (Fig. 3). It can be concluded that during liquefaction, the sediment deforms according to the vibrations [18, 19].

2.2 Vibration Propagation from the Source to the Surrounding Soil and Mechanism of Soil Densification

Liquefaction is a sudden loss of resistance occurring preferentially in granular soils under undrained and saturated conditions under dynamic motions (such as earthquakes) accompanied by a rapid increase in interstitial pressure, which could not be dissipated under the solicitation temporarily causing the dislocation of soil particles [9]. The consequences of liquefaction are its loss of shear strength, the development of large deformations, the total loss of the bearing capacity of the soil, the significant settlement, landslide, lateral movement, etc. [10].

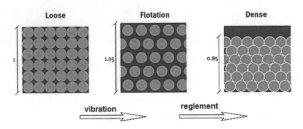

Fig. 4. The soil densification mechanism (localized liquefaction) in runs of vibroflotation treatment.

If the soil is saturated (seabed) the rearrangement goes through three stages: Loose; float and dense (Fig. 4). The overall localized liquefaction process can be separated into three major phases [16]:

- Initial arrangement little compacted: The interstitial pressure or pore pressure inside the soil is the hydrostatic pressure. The weight of the grains is carried by the granular skeleton.
- Liquefaction-consolidation (destabilization): under the effect of vibrations, the grains move and are destabilized. During a transitional period, the grains are carried by the fluid which causes an increase in pore pressure. Then the excess of interstitial pressure gradually decreases.
- Final arrangement: the grains are in contact again but in a more compact network. The pore pressure is hydrostatic and the grain weight is taken up by the granular skeleton.

As a result, the settlement is due to shear waves and wavy compression, which are transmitted from the axis of the compaction probe to the surrounding soil. The resonance compaction method uses the vibration amplification effect to increase the settlement efficiency [20].

3   Project Presentation and Soil Conditions

The purpose of the Djen-Djen port agitation study is to determine along the inner works and the harbor basin, the level of local swirling of the swell for various offshore directions for the current port configuration (Fig. 5). The agitating study showed that the development of a container terminal requires the extension of the north dike protection structures of 400 m and the east dike protection structures of 250 m with the creation of a Croin of 100 m, to reduce the width of the entrance channel. The structure of the protective works adopted being of the vertical dike (breakwater) type.

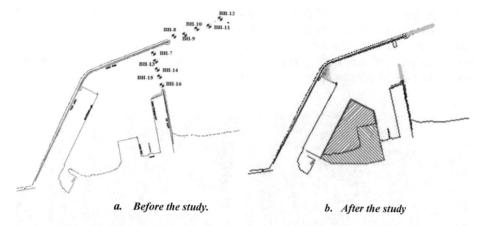

a. *Before the study.*   b. *After the study*

Fig. 5. Djen-Djen port mass plan.

3.1   Geotechnical Investigation

Drilling was conducted offshore, using a COMACCHIO (crawler) brand rig equipped with SPT testing equipment and loaded onto a floating barge (Fig. 6). Regarding the expansion of the dikes, we carried out a drilling study, a physical research and tests at the initial place on the project area; seeking to know the state of the layers on the base soil, the physical and dynamic characteristics of the soil.

Fig. 6. SPT testing equipment loaded onto a offshore floating barge.

3.1.1   Penetration (SPT) with Corer

The drilling study has the principle of advancing up to 5 m of marl using a rotating oil-pressure gauge (Fig. 7), and additional drilling is planned for the verification of the soil layer section according to bathymetric characteristics. The standard penetration test is a test method defined in NF P94 116 (ASTM D1586); this test was performed at intervals of 2.0 m. The test consists of determining the resistance to dynamic penetration of a standardized corer beaten in the bottom of a prior drilling. Depending on the depth, the sinker is given under the dead weight and the number of sheeting needed for each successive interval of 15 cm (15 cm + 15 cm) or the refusal for 50 shots of sheep for one or the other intervals. Soils that have been identified are also described. Undisturbed samples collected during standard penetration are used for soil classification and physical characteristics tests.

Fig. 7. Photography of SPT testing equipment loaded onto a offshore floating barge.

3.1.2 Soil Conditions

According to the results of the on-site study and indoor tests, the sub-soil of the study area is composed of the following order: sedimentary layer of sand, sedimentary layer of pebbles and marl (Fig. 8).

a. North pier (Jetty):

In the section of the North Dyke, five studies were carried out, drilling results of 21.5 m maximum are as follows;

- 1st layer: Loamy sand Very loose to loose,
- 2nd layer: Dense to very dense loamy sand,
- 3rd layer: Compact to very compact sandy gravel,
- 4th layer: Stiff Marne.

b. Croin on the North Pier and East Pier:

In the section of the croin and dike is, 6 hole cores; drilling results of up to 28.5 m are as follows;

- 1st layer: Loamy sand Very loose to loose,
- 2nd layer: Dense to very dense loamy sand,

(a)

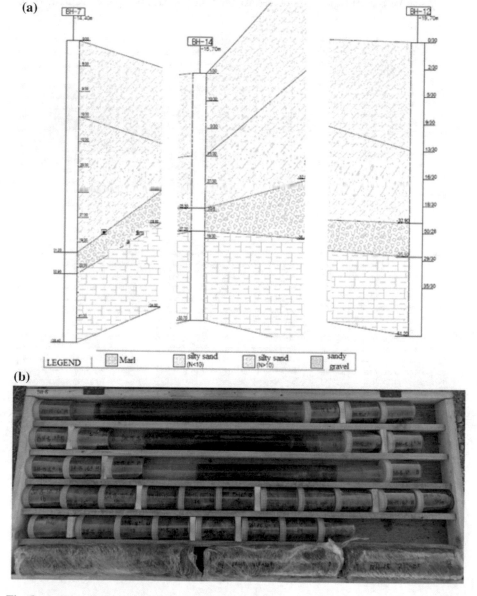

(b)

Fig. 8. (a) Lithologic section of the port site. (b) Undisturbed samples collected during standard penetration for soil classification and physical characteristics tests.

- 3rd layer: Compact to very compact sandy gravel,
- 4th layer: Stiff Marne.

3.1.3   Assessment of Liquefaction from the SPT Test

SPT test indicates liquefaction stability rate versus depth for each location. But this concerns the most unstable section for each place. calculation results for all wells are shown in Fig. 9. According to a detailed examination, the liquefaction is less than 1 stability rate below 10 N_{SPT}. It is therefore necessary measures against liquefaction.

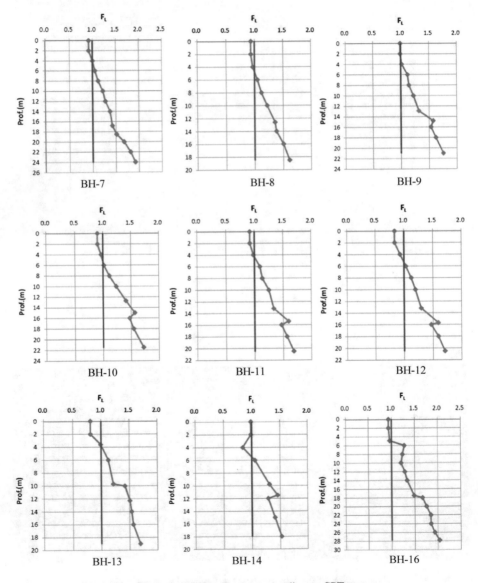

Fig. 9. Potential liquefaction according to SPT tests.

3.2 Laboratory Tests

3.2.1 Physical Tests

The sedimentary sand layer has a natural water content between 11.7 and 21.1% (average: 16.2%) and the density between 2.621 and 2.66 (average: 2.642). The amount passed through No. 200 sieve is 2.48 to 38.72%. The marl soil layer has a natural water content between 19.0 to 27.1% (average: 22.4%) and density between 2.705 to 2.727 (average: 2.717). The amount passed through the sieve No. 200 is 85.51 to 97.58%.

3.2.2 Mechanical Tests

The simple compression test and the triaxial test of undrained soil without consolidation on marly soil gave an undrained soil resistance at shear ranging from 429.1 to 509.2 kPa.

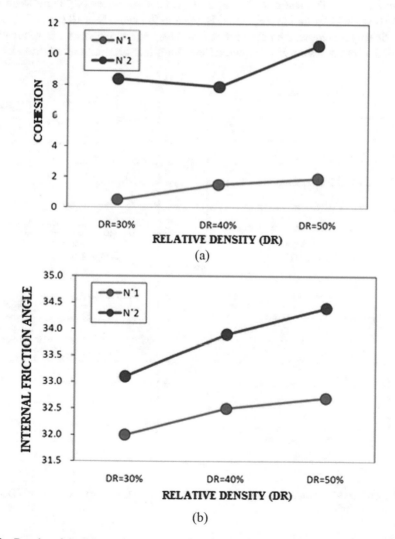

Fig. 10. Results of the Direct Shear Test by Relative Density; (a) Cohesion Vs. Dr, (b) Internal friction angle Vs. Dr.

3.2.3 Direct Shear Test According to Relative Density (ASTM D 3080-98)

The direct shear test (Fig. 10) on the sandy layer gave a cohesion between 0.5 to 8.4 kPa on a relative density of 30%, between 1.5 to 7.9 kPa on a density of 40% and between 1.9 and 10.6 kPa on a density of 50%. The internal friction angle at the relative density is 32.0 to 33.1 (deg), 32.5 to 33.9 (deg) and 32.7 to 34.4 (deg).

3.2.4 Essai Triaxial de Vibration (ASTM D 5311)

The behavior of the luminous sand with different relative density (DR-30, 50, 80) in different applied pressures, These tests were performed on consolidated samples with different cyclic loading modes with a CSR ranging from 0.25 to 0.4 and two different damping amplitudes (DA5%, DA10%). The sample under cyclic loading shows different responses of Fig. 11; witch shows the relationship of repetitive shear stress ratio to relative density. The test gave the ratio of cyclic shear stress (CSR) between 0.305 and 0.391, when DA (double amplitude) is 5% and the value N is 10 (M = 6.5). As the relative density increases, the ratio of repetitive shear stress increases. These results can be applied to the assessment of liquefaction taking into account the relative density.

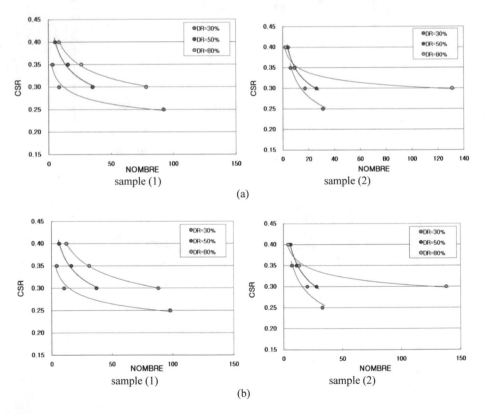

Fig. 11. Triaxial vibration test at two sample 1 and 2; CSR-N Curve (a) (DA 5%) (b) (DA 10%).

3.2.5   Results from Laboratory Examinations

The standard penetration test shows that the natural ground has sufficient resistance to liquefaction but the upper layer (first 4 m) of the lands does not meet the safety factor (FS = 1,25). The triaxial vibration test is therefore performed to calculate the cyclic

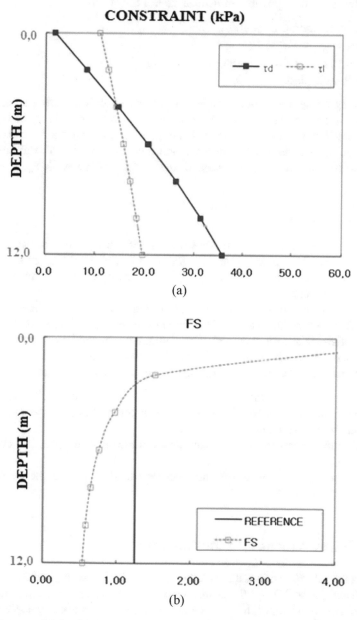

Fig. 12. The potential for liquefaction of soils from laboratory tests according to depth by the AMBRASEY method, (a) Constraint distribution, (b) Liquefaction safety factor.

resistance ratio to liquefaction. The triaxial vibration test shows the need for the liquefaction risk prevention method on the sand layer of the pier (jetty) area. If laboratory tests show the potential for soil liquefaction, then the AMBRASEY method is applied as illustrated in Fig. 12.

The results of the particle size analysis carried out for the sandy soil on the north and east dykes make it possible to deduce such a conclusion;

- Degree of saturation Sr $\fallingdotseq$ 100%,
- $2.7 < cu = \frac{D_{60}}{D_{10}} < 9.8 < 10,$
- $0.08 < 0.129 \leq$ D50 $\leq$ 0.168 m $\leq$ 2.0 mm,
- Ip = N.P (non-plastic) $\leq$ 10.

According to the results of the review of the figures above, there is a possibility of liquefaction in this works area. We therefore mentioned the possibility of liquefaction through a detailed examination. For the detailed soil liquefaction assessment, we used the results of the standard penetration test and the vibration and triaxial compression test. As for the maximum acceleration of the surface of the earth a_{max}, we applied 0.200 g.

3.3   Recommandations

On the project site there are very soft sandy soils with an N_{SPT} number of less than 10, and settlements and liquefaction risks are foreseeable. Soil improvements are required, and the vibroflottation method will be applied, in response to settlement and liquefaction that may occur.

According to the ground settlement and liquefaction test, a soil improvement will be required for the section of the dike. Taking into account the fact that this zone is composed of sandy and fabulous soil, the technique of consolidation by the vibration is recommended. Since it is on the sea, the vibroflottation technique is recommended. To apply this technique requires a detailed examination of the following;

- Careful observation of the current state of the soil; particle size, N_{SPT},
- Bearing capacity required after soil improvement; potency necessary for the stability of the constructions,
- Checking the effects of improvement during the works; test jobs and change consolidation intervals,
- Verification of effects after soil improvement; loading test on the plate, standard penetration test, PDL.

4   Soil Improvement by Vibroflottation

Regarding the vibroflotation test board. $4.0 \times 4.0 \times 4.0$ (16 holes) and $3.5 \times 3.5 \times 3.5$ (16 holes) and $3.0 \times 3.0 \times 3.0$ (16 holes) are vibroflotation test areas the test duration is 60 s per 1 m depth for each hole. When the vibroflotation test is complete, the SPT test will be done to verify the characteristics of the seabed and to ensure soil improvement in terms of density and bearing capacity in order to ensure the soil stability of caissons' foundations. If the results of the SPT are acceptable, use the same time, the same mesh and the same height for the remains.

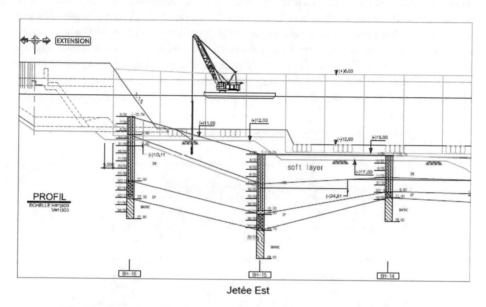

Jetée Est

Fig. 13. Drawing of the Vibroflotation in the subsoil of the East Pier [21].

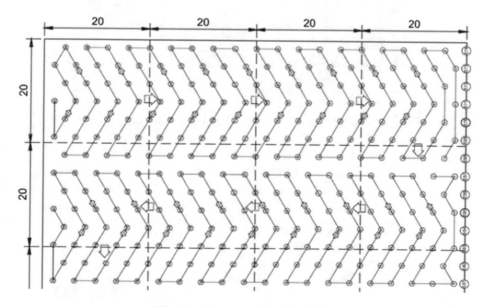

Fig. 14. Sequence - East Pier [21].

The sequence of work: Is as below who has the possibility of variable according to the conditions of the site. Using a floating crane equipped with the DGPS system to position the points.

The Vibroflotation will be done as shown in Figs. 13, 14 and 15 and in accordance with the main performance factors of vibroflotation confirmed by the test board. Nevertheless, the construction test board will be made with different dimensions of triangles, to determine the adequate mesh for a good improvement of the soil which will ensure the stability of the foundations of the caissons. The distance chosen between each hole is 3M. The latter is well verified to apply it to the existing seabed in accordance with test board. The first step is the lowering to the top of the marly layer. The second step is vibration during the designated time. The third step moves to the

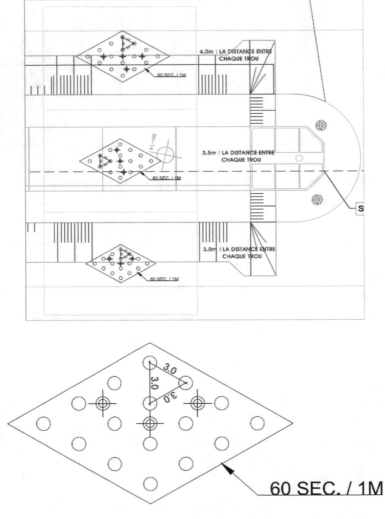

Fig. 15. Vibroflotation test at the East Pier and chosen mesh [10, 11].

next hole (point). The vibration time and the travel height will be confirmed by the vibroflotation test (Fig. 15).

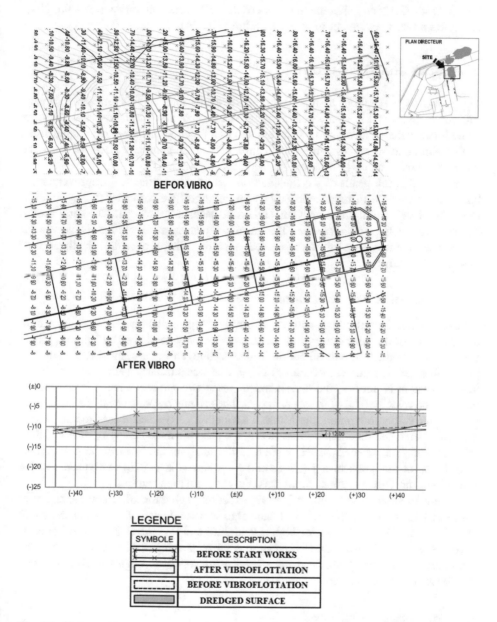

Fig. 16. Bathymetric survey results before and after vibroflotation of JETEE EST at the DjenDjen port.

With reference to the comparison with the actual bathymetric survey before the treatment of the ground of the pier (Quai) East of the port of DjenDjen carried out on January 18th, 2012, and after the treatment of the soil carried out on August 14th, 2012, one notes a mean settlement of about 87.5 cm (Fig. 16).

5   Results and Discussions

The seabed is susceptible to liquefaction settlements and, in this case, the damage suffered is considerable as the deterioration of the protective structures etc. Target N value is calculated at maximum 15 and applied as a function of depth. After application of the vibroflotation process, the densification effects are generally good for a surface layer of 4.0 m, which will ensure liquefaction safety (Fig. 17).

Figure 18 is the result of the liquefaction test assumed after soil improvement; it takes more than 45% of relative density by improving. If we convert it to N_{SPT}. we get more than 15/30. So improvement soil should be necessary during 15/30; but it should be checked by the soil engineer. The results illustrate an evolution is very clear mechanical characteristics of the soil treated in reality; What proves the effectiveness of this type of treatment.

5.1   Liquefaction Risk Assessment

Among the various drill points carried out during the project's geotechnical campaign, an evaluation of the liquefaction was carried out in sandy soils at BH-15 points. Given the soil improvement by vibroflotation in perspective, an increase in shear strength is predictable, and the number of strokes N_{SPT} tests was similarly increased. Following this increase in N strokes, the correlation of these two variables made it possible to estimate the velocity (Vs) of the shear waves. The seismic response analysis taking into account the soil amortization, the Module Reduction Curve and the Mitigation curve must be entered and the models integrated in the program must be appropriately chosen for each soil. The seismic waves applied in the analysis are the real long and short period data of the Tokachi-oki (1968) and Miyagiken-oki (1978) earthquakes in the Hachinohe and Ofunato ports. Some seismic wave data are fictitious.

Figure 19 shows the results of the liquefaction evaluation after soil treatment at points BH-15 where the geotechnical campaign was conducted. The results show that the whole soil has a safety factor higher than 1.25 and that the risk of liquefaction is lower. However, it is preferable to be cautious when carrying out the work because the soil characteristics after treatment have been estimated from empirical formulas and differences can be observed.

5.2   Monitoring of Settlement

A monthly settlement check (topographic survey) of the caisson above our actual treated soil; found an maximum of 9.00 cm of settlement; illustrated in Fig. 20. This value in displacement is due to the effect of the soil treatment (Vibroflottation), giving an increase in bearing capacity and an improvement of the compactness (density) of the

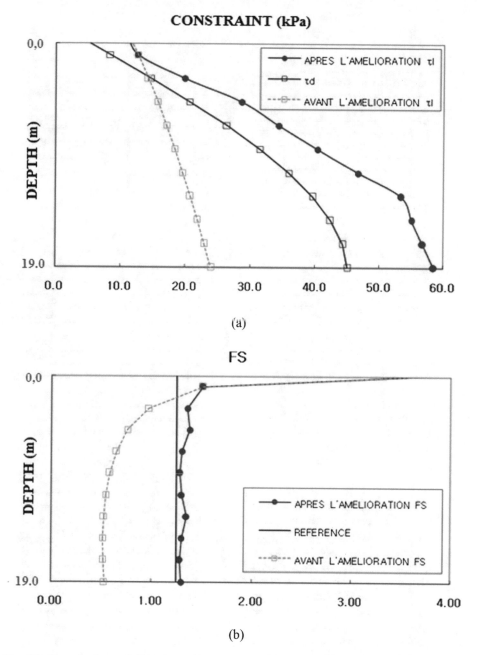

Fig. 17. Examination of liquefaction after the application of vibroflotation test board; (a) Relative density as a function of depth, (b) Liquefaction safety factor.

52   K. Houssam

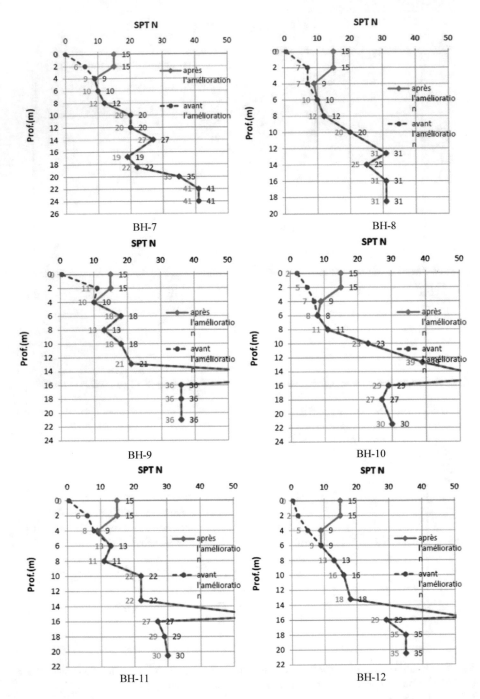

Fig. 18. NSPT before (red) and after (blue) improvement by vibroflotation.

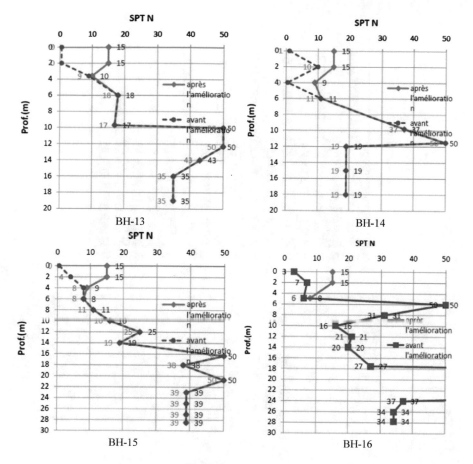

BH-13   BH-14

BH-15   BH-16

Fig. 18. (*continued*)

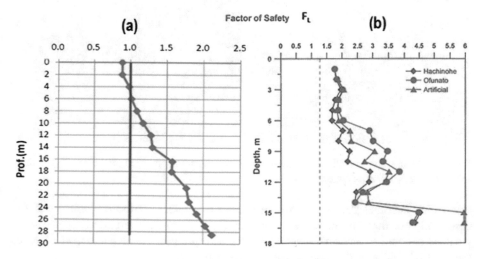

Fig. 19. Seismic response analysis to evaluate liquefaction: Liquefaction safety factor before (a) and after (b) treatment in BH-15.

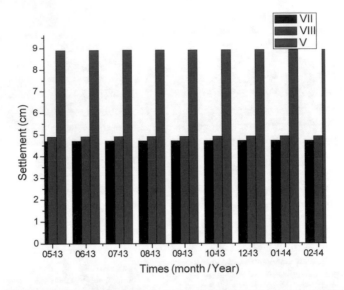

Fig. 20. Comparison of settlement curves of in-situ measurements (topographic survey) of three caissons as a function of time after soil improvement by vibroflotation.

soil which becomes denser and which has a great effect on the settlement and the deformation of the soil. Since the construction of the crown beam and its accessories, we have not noticed any settlement or geotechnical problems encountered, which gives the high reliability of this marine soil treatment method.

6   Conclusion

The vibroflotation technique takes into account the optimum compaction and homogenization of the characteristics of all the granular soil, whether embankments or sandy deposits, dry or below the water table. While the principle of compaction of vibroflotation (grains in a denser state by vibration) is a simple concept, the application of technology in an optimal way is always an art that few have mastered. This technique consists of densifying and increasing the compactness of the soil in order to improve their physical-mechanical characteristics by applying horizontal forces which cause a rearrangement of the soil grains and become denser. The degree of improvement achievable depends on the energy of the vibrator, the depth of the vibrator penetrations, the amount of time spent densifying the soil, and the mesh size.

The treatment of soils by vibroflotation has taken a scale in Algeria in recent years, it has been applied at the level of the DjenDjen port in Jijel province, object of our study, as part of its extension, to improve the soil support that will receive the foundations of protective structures. The goal is to understand and understand the method of execution of this technique, its mechanism during treatment, and their effect on the behavior of the soil during and after its implementation. The effectiveness of this method of soil treatment was demonstrated by the results of bathymetric and topographic surveys as well as the in-situ tests available including the SPT tests which made it possible to check the capacity of the support soil (scabed) before and after realization of vibroflotation. In addition, this treatment to minimize the risk of liquefaction and instability of the protective structure, in addition to the advantage of speed of execution and reasonable cost compared to the importance of the project, thus no adverse effect has been reported on the environment.

Although vibroflotation gives very satisfactory results for soil improvement, to our knowledge there is no scientific descriptive, experimental or analytical research on this method and its effect on the soil of interaction, apart from the references. As perspectives in this research theme, it is recommended to study several parameters influencing treatment outcomes, such as the optimal distance between the points of vibroflotation, the threshold of the depth to be treated, and how to stop the foreseeable settlement at the end of treatment and creation of approaches to prevent improved mechanical parameters.

References

1. Bell, F.G., Detry, V.: Méthode de traitement des sols instables. Eyrolles, Paris (1978)
2. Farhat, H., Robert, J., Berthelot, P.: Extension du port de la Condamine à Monaco - Confortement des sols en place et des remblais sous-marins. Rev. Fr. Geotech. **112**, 29–34 (2005)
3. Lehuérou-Kérisel, J., Caquot, A., Kersiel, J.: Traité de la mécanique des sols, 4eme édition, Paris (1966)
4. Sreekantiah, H.R.: Vibroflottation for ground improvement - a case study. In: Proceedings of the Third International Conference on Case Histories in Geotechnical Engineering, St. Louis, Missouri, pp. 949–954 (1993)

5. Khelalfa, H.: Traitement du sol par vibroflottation application aux ouvrages de protection du port de DjenDjen. Jijel Algérie. J. Mater. Eng. Struct. **3**(4), 149–160 (2016)
6. Amélioration des sols (Vibroflottation) Ménard, Doc. Keller (1974)
7. Sayar, A.D., Khalilpasha, M.: Soil improvement using vibro replacement technique. Int. J. Adv. Environ. Biol. **6**(2), 658–661 (2012)
8. McCabe, B.A., McNeill, J.A., Black, J.A.: Ground improvement using the vibrostone column technique. In: Proceeding of the Meeting of Engineers Ireland West Region, NUI Galway, 15 March 2007
9. Jefferies, M., Been, K.: Soil Liquefaction, a Critical State Approach. CRC Press, Boca Raton (2015)
10. Khelalfa, H.: Traitement du sol par vibroflottation, I.: Proceeding of 2eme Séminaire national sur les géo-risques, Université Mohammed Seddik Benyahia- Jijel-Algérie, 17 et 18 Novembre 2015
11. Khelalfa, H.: Traitement du sol par vibroflottation. In: Proceeding of 3ème communication Journées d'Etudes CGCE, Université Mohammed Seddik Benyahia- Jijel-Algérie 13–14 mai 2014
12. Tarzaghi, K., Peck, R.B., Mesri, G.: Soil Mechanics in Engineering Practice, 3rd edn. Wiley, New York (1996)
13. Mecsi, J., Gökalp, A., Düzceer, R.: Compactage des remblais hydrauliques par la technique de vibroflotation. In: Proceeding of the 16th International Conference on Soil Mechanics and Geotechnical Engineering, Université de Pécs, Hungary (2005)
14. Truong, P.H.V.: Dynamic excess pore water pressures by dynamic soil masses and dynamic water heights. Int. J. Geol. **3**(6), 77–83 (2012)
15. Gökalp, A., Düzceer, R.: Vibratory deep compaction of hydraulic fills. In: Proceeding of the XIIIth European Conference on Soil Mechanics and Geotechnical Engineering, ISSMGE, Prague, Czech Republic (2003)
16. Aussillous, P., Collart, D., Pouliquen, O.: Liquéfaction des sols sous vagues. In: Proceeding of 18ème Congrès Français de Mécanique Grenoble, 27–31 août 2007
17. Andrus, R.D., Chung, R.M.: Ground Improvement Techniques for Liquefaction Remediation Near Existing Lifelines. Report, NISTIR 5714, Building and Fire Research Laboratory, National Institute of Standards and Technology, Gaithersburg (1995)
18. Giese, S.: Numerical Simulation of vibroflotation compaction – Application of dynamic boundary conditions. In: Konietzky (ed.) Numerical Modeling in Micromechanics via Particle Methods (2003)
19. Vernay, M., Morvan, M., Breul, P.: Etude du comportement des sols non saturés à la liquéfaction. In: Proceeding of 33èmes Rencontres de l'AUGC, ISABTP/UPPA, Anglet, 27 au 29 mai 2015
20. Massarsch, K.R.: Effects of vibratory compaction. In: International Conference on Vibratory Pile Driving and Deep Soil Compaction, Louvain-la-Neuve, Keynote Lecture, pp. 33–42 (2002)
21. Khelalfa, H.: Coastal Soil Treatment to Stabilize Vertical Breakwaters. LAMBERT Academic Publishing (2018). ISBN: 978-3-330-03812-7

Enhancing the Service Life of Aged Asphalt Concrete by Micro Crack Healing and Recycling

Saad Issa Sarsam[✉] and Mostafa Shaker Mahdi

Department of Civil Engineering, College of Engineering,
University of Baghdad, Baghdad, Iraq
saadisasarsam@coeng.uobaghdad.edu.iq

Abstract. Aging of the asphalt cement in the pavement changes the quality of asphalt concrete from flexible to stiff and the pavement will be susceptible to all types of distresses. Recycling is considered as a good alternative to enhance the pavement for additional service life. Recycling agent can provide the required flexibility and increase its micro crack healing potential. In this investigation, aged asphalt concrete was recycled with two types of additives, carbon black CB and styrene Butadiene rubber SBR. Two set of Cylindrical specimens have been prepared, the first set has 102 mm diameter and 63.5 mm height while the second set has 102 mm diameter and 102 mm height. The first set was subjected to repeated indirect tensile stresses (ITS) at 25 °C and tested under stress level of 138 kPa, while the second set practices repeated compressive stresses (CS) at 40 °C and tested under three stress levels of (69, 138, and 207) kPa. All the specimens were tested in the pneumatic repeated load system PRLS and constant loading frequency of 60 cycles per minute. The loading sequence for each cycle is 0.1 s of load duration and 0.9 s of rest period. After 1000 load repetitions which allowed for the initiation of micro cracks, the test was terminated. Specimens were stored in an oven for 120 min at 60 °C to allow for crack healing by external heating. Specimens were returned to the PRLS chamber and subjected to another cycle of stresses repetition. The impact of crack healing was measured in terms of the change in Resilient Modulus Mr under ITS and permanent deformation under CS before and after healing for each recycling agent. It was concluded that Mr under repeated ITS increases by (25, 30, and 20) % for aged, CB treated and SBR treated mixtures respectively after healing. On the other hand, the permanent deformation under repeated CS decreases by (31, 43, and 45) %, (6, 49, and 10.6), (19, 24.5, and 13.2) % for aged, CB treated and SBR treated mixtures under stress levels of (69, 138, and 207) kPa respectively after healing.

Keywords: Recycling · Healing · Microcrack · Deformation · Resilient modulus · Asphalt concrete

1 Introduction

Influence of excessive loading and environmental impact on asphalt concrete are considered as major issues for initiation of microcracks in the pavement structure. The desire to discover more sustainable paving practices are forcing agencies to look for

© Springer Nature Switzerland AG 2020
H. Shehata et al. (Eds.): GeoMEast 2019, SUCI, pp. 57–69, 2020.
https://doi.org/10.1007/978-3-030-34184-8_3

routes for augmenting the re-utilization of reclaimed asphalt pavement (RAP), (Zaumanis and Mallick 2015). While the greater part of the academic and industrial establishments has been focused on the development of techniques to reuse HMA with up to 40% RAP content, a few industry innovators have superior 100% reusing advances in the recent four decades to a degree where routine production of 100% reused mixes is in clear vision, (Zaumanis et al. 2016). The primary obstruction in the widespread utilization of 100% reusing is the dubious performance of 100% RAP and absence of a unified and objective framework for choice of materials and mix design. (Sarsam 2016) stated that the Asphalt concrete mixture is considered to have nonlinear viscoelastic behavior, its fatigue life consists of two components, namely the resistance to fracture and crack, and the ability to heal the micro cracks. Both processes change with temperature and time. Such processes exhibit the sustainability potential of asphalt concrete pavement. As stated by (Al-Qadi et al. 2007), after more than 30 years since its first trial in Nevada and Texas, it appears that the use of RAP will not only be a beneficial alternative in the future but will also become a necessity to ensure economic competitiveness of flexible pavement construction. The use of rejuvenators has the potential to restore rheology and chemical components of aged RAP bitumen, thus allowing a significant increase in the amount of RAP to be properly implemented in hot mix asphalt HMA. Results show that rejuvenators modify bitumen chemistry and consequently rheology by enhancing the viscous response, (Mazzoni et al. 2018). The dynamic behavior of the recycled asphalt concrete (with cutback and emulsion) in terms of the resilient modulus (Mr), rutting resistance, and permanent microstrain have been investigated by (Sarsam and Saleem 2018a). It was concluded that RAP mixture can hold the applied loading with minimal permanent deformation as compared to the recycled mixtures. The resilient modulus is lower by (24 and 39) % for mixes recycled with cutback and emulsion respectively as compared to that of RAP. The rate of strain (slope) increases by 11% and 4% when cutback and emulsion when implemented as a recycling agent respectively as compared to that for RAP mixture. (Pradyumna and Jain 2016) describes the comparison of properties of mixture with recycling agents, which has been prepared in laboratory on the RAP material, and their performance has been compared with virgin mixes. Various performance tests such as Retained Stability, Indirect Tensile Strength (ITS) and Tensile Strength Ratio (TSR), and Resilient Modulus test has been carried out to compare the performance of RAP modified mixes and virgin mixes. It was concluded that the laboratory results indicate that the bituminous mixes with RAP and recycling agent provide better performance compared to virgin mixes. (Sarsam 2007) investigated the recycling of asphalt concrete and concluded that aging increases Hveem cohesion, but it has negative effects on Marshall, tensile, and flexural properties. Recycling has a positive effect on Asphalt concrete overall properties and have changed the mode of deflection from cracking to bending. The variation in gradation exhibits significant effects on both aged and recycled conditions of Asphalt concrete. (Mhlongo et al. 2014) examines the use 100% reclaimed asphalt pavement (RAP) for sustainable construction and rehabilitation of roads. The recovered aggregates fall within the envelope for continuously graded mix and the recovered binder is 5.3%. The new hot mix asphalt was design with virgin softer bitumen grade 50/70 to act as rejuvenator at 0, 0.3% and 0.6% to RAP. Test such as Marshall test, indirect tensile strength (ITS), fatigue resistance and workability were conducted. ITS meets the minimum specified for all the percentages of bitumen. Fatigue resistance of the recycled mixtures increases as the bitumen content increases. It was concluded that the performance of 100% RAP in terms of air voids, stability and ITS at 5.9% binder will be a good material for road construction and rehabilitation. (Sabhafer and Hossain 2017) investigated the effects of rejuvenation on hot in place recycling HIR performance by assessing critical performance indicators such as cracking resistance, moisture susceptibility, and low temperature cracking. An experimental program was designed that included mechanical property measurements of the HIR mixture by conducting Texas overlay, thermal stress restrained specimen, and moisture susceptibility tests. Study results showed significant variability in the mechanical performance of HIR mixtures, which was attributed to the variability of binders.

(Sarsam and Husain 2017) investigated the influence of healing cycles and asphalt content on resilient modulus of asphalt concrete. It was concluded that permanent deformation decreases as the healing cycles increase. Mr under indirect tensile stress (ITS) increases by (33.4, 100), 100, and (25, 150) % after one and two healing cycles respectively as compared with control mix for mixes with 4.4, 4.9, and 5.4% asphalt content. (Silva et al. 2012) investigated if totally recycled HMA mixtures could be a good solution for road paving, by evaluating the merit of some rejuvenator agents (commercial product, used engine oil) in improving the aged binders' properties and the recycled mixture performance. Totally recycled HMAs were produced and their performance (water sensitivity, rutting resistance, stiffness, fatigue resistance, binder aging) was assessed. It was concluded that totally recycled HMAs can be a good alternative for road paving, especially if rejuvenator agents are used to reduce their production temperature and to improve their performance. (Valdés et al. 2011) presented an experimental study to characterize the mechanical behavior of bituminous mixtures containing high rates of reclaimed asphalt pavement (RAP). Two semi-dense mixtures containing 40% and 60% RAP, respectively were used for rehabilitation of a highway section. The mechanical properties were studied by determining the stiffness modulus and indirect tensile strength and cracking and fatigue behavior. Results show that high rates of recycled material can generally be incorporated into bituminous mixes by proper characterization and handling of RAP stockpiles. The aim of the present investigation is to verify the influence of two types of recycling agent and the crack healing concept on the resilient modulus and permanent deformation of aged asphalt concrete.

2 Materials and Methods

Materials implemented in this research are locally available, and economically valuable. They could be categorized into three groups, aged (reclaimed) asphalt concrete mixture, asphalt cement, and recycling agents.

2.1   Aged Asphalt Concrete Mixture

The aged asphalt concrete mixture was obtained by the rubblization of the asphalt concrete binder course layer from highway section at Karbala province. This highway was constructed in 2012 and suffers the deformation due to high traffic loads. The aged asphalt mixture obtained was free from dust and loam that may be stick on the top surface. The aged mixture was heated to 130 °C, combined and reduced to testing size; a representative sample was exposed to Ignition test based on (AASHTO 2013) procedure to obtain binder and filler content, gradation and properties of aggregate. Table 1 illustrates the properties of aged mixture. On the other hand, the properties of the extracted coarse and fine aggregates and mineral filler from the aged asphalt concrete mixture are demonstrated in Table 2. The gradation of the aged asphalt concrete mixture after ignition test is shown in Fig. 1. It can be observed that the gradation of aggregate is within the Specification limits of Roads and Bridges (SCRB 2003) for binder course layer.

Table 1. Properties of aged mixture obtained after ignition test

Material	Property		Value
Asphalt binder	Binder content %		3.84
Aged mixture	Marshall properties	Stability	17.532 KN
		Flow	2.9 mm
		Bulk density	2.320 gm/cm^3
		Air voids	5.1%
		Theoretical maximum density	2.448 gm/cm^3

Table 2. Properties of coarse and fine aggregates and mineral filler extracted from aged asphalt concrete

Material	Property	Value
Coarse aggregate	Bulk specific gravity	2.62
	Apparent specific gravity	2.76
	Water absorption %	1.021
	Percent of fracture faces %	93
Fine aggregate	Bulk specific gravity	2.67
	Apparent specific gravity	2.81
	Water absorption %	1.82
Mineral filler	Percent passing sieve no. 200	98
	Specific gravity	3.15

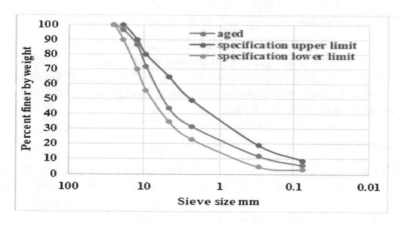

Fig. 1. Gradation of aged asphalt concrete mixture

2.2 Additives

Two types of additives namely carbon black and Styrene Butadiene Rubber (SBR) have been implemented in this work to prepare the recycling agent. Three percentages of additives (0.5, 1, and 1.5) % by weight of asphalt cement and two percentages of asphalt cement (1 and 2) % by weight of mixture have been blended and implemented as reju-venator and mixed with the aged asphalt concrete. Table 3 shows the properties of carbon black, while Table 4 present the properties of SBR as supplied by the manufacturer.

Table 3. Properties of carbon black

Property	(ASTM 2009)	Test result
Residue on sieve no. 35	D-1514	10
Pour density gm/liter	D-1513	352.4
Ash content %	D-1506	0.75
PH	D-1512	7.5–9
Specific surface area (m^2/g)	D-6556	36
Oil absorption number	D-2414	122
Particle diameters (nanometers)	–	120

Table 4. Properties of SBR

Property	Value
Specific gravity (g/cm^3 at 25 °C)	1.01
Color	Milky, white, liquid
Chloride content	Nil
Butadiene (% by wt.)	40
Mean part size (micro-nicle)	0.17
Viscosity	Low

2.3   Recycling Agents

Two types of recycling agent have been selected and prepared in the laboratory based on the available literature and previous investigations, (Sarsam and AL-Zubaidi 2014), (Sarsam and AL-Shujairy 2015), (Sarsam and Mahdi 2019) and implemented in this work. They are (asphalt cement mixed with carbon black) and (asphalt cement mixed with Styrene Butadiene Rubber SBR).

2.4   Asphalt Cement

Asphalt cement of penetration grade (40–50) was obtained from Al-Dura refinery and implemented for the recycling process. Asphalt cement testing confirmed that its properties conform to the specifications of State commission for Roads and Bridges (SCRB 2003). Its physical properties are listed in Table 5.

Table 5. Physical properties of asphalt cement

Physical properties	ASTM designation	Asphalt cement	(SCRB 2003)
Penetration	D5-06	43	40–50
Softening point	D36-95	46	–
Ductility	D113-99	140	>100
Specific gravity	D70	1.04	
Flash point	D92-05	269	>232
Retained penetration of residue	D5-06	57%	>55
Ductility of residue	D113-99	73 cm	>25

2.5   Blending of Asphalt Cement with Carbon Black and Styrene Butadiene Rubber SBR

Asphalt cement of penetration grade (40–50) from Al-Dura refinery was mixed with 1.5% of carbon black (by weight of added asphalt) which was obtained from local market in powder form. Asphalt cement was heated to approximately 130 °C, and the carbon black was added steadily to the asphalt cement and mixed until homogenous blend was accomplished. The mixing was sustained for thirty minutes by a mechanical blender. On the other hand, Asphalt cement of penetration grade (40–50) from Al-Dura refinery was mixed with 1.5% of SBR (by weight of added asphalt) which was obtained from local market in liquid form. Asphalt cement was heated to approximately 130 °C, and the SBR was added steadily to the asphalt cement with thrilling until homogenous blend was accomplished, the mixing and thrilling were sustained for thirty minutes by a mechanical blender.

2.6   Preparation of Recycled Asphalt Concrete Mixtures and Specimens

The Aged mixture which was obtained from the reclaimed material from the site. The specimens were prepared to explore the performance after recycling. Recycled mixture consists of 100% aged asphalt concrete and rejuvenator (recycling agent) blended together at specified percentages depending on the mixing ratio. Aged asphalt concrete was heated to 160 °C while the rejuvenator was heated to 120 °C before it was added to the aged mixture. The rejuvenator was added as a percentage of asphalt content and mixed for two minutes until all mixture was visually covered with recycling agent as addressed by (Sarsam 2007). The recycled mixture was prepared using two types of recycling agents, asphalt cement mixed with carbon black and asphalt cement mixed with SBR. Two types of specimens have been prepared, the first type is a standard Marshal specimen of 102 mm in diameter and 63.5 mm in height while the second type has 102 mm diameter and 102 mm height. For the first type, the asphalt mixture was placed in the preheated mold, and then it was spaded with a heated spatula 15 times round the perimeter and 10 times round the inner. The temperature of mixture directly prior to compaction temperature was (150 °C). The mold was fixed on the compaction pedestal and (75) blows on the top and the bottom of specimen were applied with identified Marshal compaction hammer. On the other hand, the second type of specimens was constructed under static compaction to the same target density of the Marshal specimen using the same compaction temperature. The mixture was compacted at temperature of (150 °C) under an initial load of one MPa to set the mixture against the sides of the mold. After that, the required load of 20 MPa was applied for two minutes. The specimens in the mold were left to cool at room temperature for one day, then were removed from the mold using mechanical jack. Specimens of aged asphalt concrete before recycling were also prepared for comparison.

2.7   Repeated Indirect Tensile Stress Test

Specimens of the first type (Marshal) were subjected to the repeated indirect tensile stresses according to the procedure of (ASTM 2009). In this test, the specimen was stored at room temperature of 25 °C for one day; then the specimen was fixed on the vertical diametrical level between the two parallel loading bands (12.7 mm) in wide. The specimen was fixed in the pneumatic repeated Load system apparatus (PRLS) shown in Fig. 2. Asphalt concrete specimens were subjected to repeated indirect tensile stress for 1000 load repetitions at 25 °C to allow the initiation of micro cracks. Such timing and test conditions were suggested by (Sarsam and Saleem 2018b) and (Sarsam and Mahdi 2019). Such load assembly applies indirect tensile stress on the specimen in the form of rectangular wave with constant loading frequency of (60) cycles per minutes. A heavier sine pulse of (0.1) sec load duration and (0.9) sec rest period was applied over the test duration. Before the test, dial gage of the deformation reading was set to zero and the pressure actuator was adjusted to the specific stress level equal to 0.138 MPa. A digital video camera was fixed on the top surface of the (PRLS) to capture dial gage reading. The average deformation of duplicate specimens was calculated and considered for obtaining the resilient modulus.

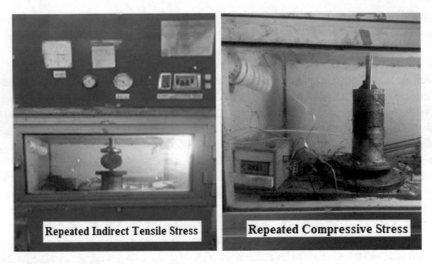

Fig. 2. Repeated stress test in progress in the PRLS

2.8   Repeated Compressive Stress Test

Specimens of the second type were subjected to the repeated compressive stresses according to the procedure of (ASTM 2009). In this test, the specimen was stored at room temperature of 25 °C for one day; then the specimen was fixed in the pneumatic repeated Load system apparatus (PRLS) shown in Fig. 2. Asphalt concrete specimens were subjected to repeated compressive stress for 1000 load repetitions at 40 °C to allow the initiation of micro cracks. Such timing and test conditions were suggested by (Sarsam and Saleem 2018b) and (Sarsam and Mahdi 2019). The load assembly applies compressive stress on the specimen in the form of rectangular wave with constant loading frequency of (60) cycles per minutes. A heavier sine pulse of (0.1) sec load duration and (0.9) sec rest period was applied over the test duration. Three types of stress level were implemented (0.068, 0.138 and 0.206) MPa. Before the test, dial gage of the deformation reading was set to zero and the pressure actuator was adjusted to apply the specific stress level. A digital video camera was fixed on the top surface of the (PRLS) to capture dial gage reading. The average deformation of duplicate specimens was calculated and considered for analysis.

2.9   Crack Healing Technique

Crack Healing technique adopted in this work was healing by the external heating. After 1000 load repetitions, to allow for the initiation of micro cracks, the test was terminated. Specimens were withdrawn from the testing chamber and stored in an oven for 120 min at 60 °C to allow for crack healing as recommended by (Sarsam and Saleem 2018a), (Sarsam and Saleem 2018b), (Sarsam and Mahdi 2019). Healing occurred in the asphalt concrete mixture specimens due to the reduction in the viscosity of asphalt cement due to external heating. The specimens were cooled at room temperature for 24 h, it was conditioned by placing in the PRLS chamber at temperature

(25 °C) for 120 min, then the specimens were subjected to another 1000 load repetitions of indirect tensile or compressive stresses.

2.10 Healing Indicators

The healing of asphalt concrete is usually evaluated by the recovery of the material's mechanical properties, such as increment in tensile strength and resilient modulus, reduction in permanent deformation and rate of deformation which have been considered as healing indicators by (Sarsam and Saleem 2018b), and (Sarsam and Mahdi 2019). The commonly used healing index is the ratio of the material strength or properties after healing to the original strength or properties. In this case, a higher ratio (more than one) indicates a better healing performance for strength evaluation, while lower ratio (lower than one) indicates a better resistance to deformation. In this investigation, the healing index obtained from the recovery of resilient modulus and reduction of deformation were used to measure the healing ability of asphalt concrete.

3 Results and Discussions

3.1 Effect of Crack Healing on Resilient Modulus Under ITS

Figure 3 illustrations the effect of the crack healing technique on (Mr) under ITS for aged and recycled mixture (unconditioned and conditioned) when subjected to repeated level of stress of (0.138 MPa). Table 6 demonstrates the healing indicators of resilient modulus (Mr) of asphalt concrete after 1000 load repetitions. It can be observed that the (Mr) increases after crack healing cycle for aged and recycled mixture. The proportion of increase was (25, 30, 20) % for aged and recycled mixture with (carbon black-asphalt and SBR-asphalt rejuvenators) respectively. This could be attributed as a result of the healing of the cracks in addition to the evaporation of more volatiles and thus increases the Resilient Modulus. Such behavior agrees well with the work reported by (Sarsam and Husain 2017) and (Shunyashree et al. 2013).

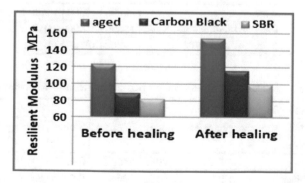

Fig. 3. Resilient modulus (Mr) before and after healing cycle under (ITS).

3.2   Effect of Crack Healing on Permanent Deformation Under Repeated Compressive Stress

Table 7 exhibit the permanent deformation of aged and recycled asphalt concrete under three levels of repeated compressive stresses. It can be observed that the deformation increases after recycling the aged asphalt concrete. This could be attributed to the added flexibility and reduced viscosity of the aged binder after digestion with the rejuvenators. On the other hand, carbon black-asphalt rejuvenator exhibits more flexibility and higher deformation regardless of the stress applied as compared with aged or SBR-asphalt recycled mixture. Table 8 exhibit the influence of recycling on the permanent deformation parameters. The first parameter is the intercept which represents the permanent strain at N = 1 (N is the number of load cycles), the higher the value of the intercept, the larger is the strain and the potential of permanent deformation. It can be observed that the intercept increases as the stress level application increase. It can be noted that recycling process has increases the intercept indicating more flexibility gained as compared to the aged mixture. On the other hand, the second parameter is the slope which represents the rate of change in the permanent strain as a function of the change in loading cycles (N) in the log-log scale. High slope value of the mix indicates an increase in material deformation rate, hence, less resistance against rutting. It can be noted that recycling process has almost increases the slope due to reduced viscosity of

Table 6. Healing indicator of resilient modulus under repeated (ITS)

Mixture type	Healing index of resilient modulus
Aged	1.25
Recycled with carbon black	1.29
Recycled with SBR	1.20

Table 7. Permanent microstrain after 1000 loading cycles of compressive stress

Mixture type	Stress level (MPa)		
	0.068	0.138	0.206
Aged	8800	10300	14400
Recycled with carbon black	14600	25400	25800
Recycled with SBR	9200	13100	15200

Table 8. Influence of stress level on permanent deformation parameters

Stress level MPa	0.068		0.138		0.206	
Deformation parameter Mixture type	Intercept	Slope	Intercept	Slope	Intercept	Slope
Aged	277.31	0.4966	556.73	0.4227	648.9	0.4218
Carbon black	199.98	0.6152	466.95	0.5655	828.64	0.5036
SBR	402.29	0.4304	614.26	0.407	1669	0.3159

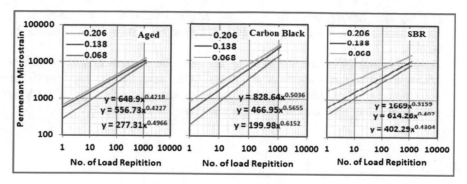

Fig. 4. Influence of recycling agent on deformation of asphalt concrete

Table 9. Healing indicator of permanent (Microstrain) under repeated (CS)

Mixture type	Stress level (MPa)		
	0.068	0.138	0.206
Aged	0.68	0.94	0.80
Recycled with carbon black	0.56	0.50	0.75
Recycled with SBR	0.55	0.89	0.86

the binder. Figure 4 demonstrations the influence of recycling agent on permanent deformation. Table 9 exhibit the impact of healing cycle after (1000) load repetitions of compressive stress (0.068, 0.138, and 0.206) MPa on permanent deformation. It can be noted that the micro-strain after the healing cycle decline. The deformation declines by (31.8, 5.8 and 19)%, (43,49 and 24)% and (44,10 and 13)% under three level of compressive stresses (0.068, 0.138, and 0.206) MPa at (40 °C) after one healing cycle for old and recycled mixture with (carbon black-asphalt and SBR-asphalt) respectively when compared with deformation before (healing cycle). The effect of (carbon black-asphalt) rejuvenator on strain is significantly visible when compared with old and (SBR-asphalt rejuvenator). Such performance completely agrees with (Sarsam and Jasim 2018).

4 Conclusions

Based on the limitations of the testing program, the following conclusions are drawn

1- Resilient modulus (Mr) after (1000) load repetitions of indirect tensile stress ITS at 25 °C increases by (25, 30.2, and 20.5)% for aged and recycled mixtures with carbon black-asphalt and SBR-asphalt respectively after healing cycle when compared with the similar mixture before healing.

2- The deformation declines by (31.8, 5.8 and 19)%, (43,49 and 24)% and (44,10 and 13)% under three level of compressive stresses (0.068, 0.138, and 0.206) MPa at

68   S. I. Sarsam and M. S. Mahdi

(40 °C) after healing cycle for aged and recycled mixture with (carbon black-asphalt and SBR-asphalt) respectively when compared with deformation before healing.

3- Healing indicators exhibit the ratio of the material strength or properties after healing to the original strength or properties. In this case, a higher ratio (more than one) indicates a better healing performance for strength evaluation, while lower ratio (lower than one) indicates a better resistance to deformation.

References

Zaumanis, M., Mallick, R.: Review of very high content reclaimed asphalt use in plant-produced pavements: state of the art. Int. J. Pavement Eng. **16**(1), 39–55 (2015). https://doi.org/10.1080/10298436.2014.893331

Zaumanis, M., Mallick, R., Frank, R.: 100% hot mix asphalt recycling: challenges and benefits. Transp. Res. Procedia **14**, 3493–3502 (2016)

Sarsam, S.I.: Sustainability of asphalt pavement in terms of crack healing phenomena: a review. Trends Transp. Eng. Appl. STM J. **3**(2), 38–55 (2016)

Al-Qadi, I., Elseifi, M., Carpenter, S.: Reclaimed Asphalt Pavement-A Literature Review. Report No. FHWA-ICT-07-001, Illinois Center for Transportation, Rantoul, IL (2007)

Mazzoni, G., Bocci, E., Canestrari, F.: Influence of rejuvenators on bitumen ageing in hot recycled asphalt mixtures. J. Traffic Transp. Eng. (Eng. Ed.) **5**(3), 157–168 (2018). https://doi.org/10.1016/j.jtte.2018.01.001

Sarsam, S., Saleem, M.: Dynamic behavior of recycled asphalt concrete. STM J. Trends Transp. Eng. Appl. **5**(3), 1–7 (2018a)

Pradyumna, T.A., Jain, P.K.: Use of RAP stabilized by hot mix recycling agents in bituminous road construction. Transp. Res. Procedia **17**, 460–467 (2016). https://doi.org/10.1016/j.trpro.2016.11.090

Sarsam, S.I.: A study on aging and recycling of asphalt concrete pavement. University of Sharjah. J. Pure Appl. Sci. **4**(2), 79–93 (2007)

Mhlongo, S.M., Abiola, O.S., Ndambuki, J.M., Kupolati, W.K.: Use of recycled asphalt materials for sustainable construction and rehabilitation of roads. In: International Conference on Biological, Civil and Environmental Engineering (BCEE-2014), Dubai (UAE), 17–18 March 2014. http://dx.doi.org/10.15242/IICBE.C0314157

Sabhafer, N., Hossain, M.: Effect of asphalt rejuvenating agent on cracking properties of aged reclaimed asphalt pavement. In: ASCE, Airfield and Highway Pavements, pp. 201–214 (2017)

Sarsam, S.I., Husain, H.K.: Influence of healing cycles and asphalt content on resilient modulus of asphalt concrete. Trends Transp. Eng. Appl. STM J. **4**(1), 23–30 (2017)

Silva, H., Oliveira, J., Jesus, C.: Are totally recycled hot mix asphalts a sustainable alternative for road paving? Resour. Conserv. Recycl. **60**, 38–48 (2012). https://doi.org/10.1016/j.resconrec.2011.11.013

Valdés, G., Pérez-Jiménez, F., Miro, R., Martenez, A., Botella, R.: Experimental study of recycled asphalt mixtures with high percentages of reclaimed asphalt pavement (RAP). Constr. Build. Mater. **25**, 1289–1297 (2011). https://doi.org/10.1016/j.conbuildmat.2010.09.016

AASHTO. Standard Specification for Transportation Materials and Methods of Sampling and Testing, American Association of State Highway and Transportation Officials, 14th Edition, Part II, Washington, D.C (2013)

SCRB. General Specification for Roads and Bridges. Section R/9, Hot-Mix Asphalt Concrete Pavement, Revised Edition. State Corporation of Roads and Bridges, Ministry of Housing and Construction, Republic of Iraq (2003)

Sarsam, S., AL-Zubaidi, I.L.: Assessing Tensile and Shear Properties of Aged and Recycled Sustainable Pavement, vol. 2, no. 9, pp. 0444–0452. IJSR Publication (2014). https://doi.org/10.12983/ijsrk-2014

Sarsam, S.I., AL-Shujairy, A.M.: Assessing tensile and shear properties of recycled sustainable asphalt pavement. J. Eng. **21**(6), 146–161 (2015)

Sarsam, S.I., Mahdi, M.S.: Assessing the rejuvenate requirements for asphalt concrete recycling. Int. J. Mater. Chem. Phys. **5**(1), 1–12 (2019)

ASTM. Road and Paving Materials, Annual Book of ASTM Standards, Volume 04.03, American Society for Testing and Materials, USA (2009)

Sarsam, S., Saleem, M.: Influence of micro crack healing on flexibility of recycled asphalt concrete. J. Adv. Civil Eng. Constr. Mater. **1**(1), 26–39 (2018b)

Shunyashree, S., Bhavimane, T., Archana, M., Amarnath, M.: Effect of use of recycled materials on indirect tensile strength of asphalt concrete mixes. IJRET: Int. J. Res. Eng. Technol. (2013). https://doi.org/10.15623/ijret.2013.0213040

Sarsam, S.I., Jasim, S.A.: Assessing the impact of polymer additives on deformation and crack healing of asphalt concrete subjected to repeated compressive stress. In: Proceedings, 17th Annual International Conference on: Asphalt, Pavement Engineering and Infrastructure, 2018 LJMU, Liverpool, UK, 21–22 February 2018

Impact of Aeration and Curing Periods on Shear Strength of Asphalt Stabilized Soil

Saad Issa Sarsam[✉] and Ahmad Zuhair Al Sandok

Department of Civil Engineering, College of Engineering,
University of Baghdad, Baghdad, Iraq
saadisasarsam@coeng.uobaghdad.edu.iq

Abstract. The subgrade soil is the foundation for the roadway structure, its responsibility is to provide the required stability for the overlaying pavement and to limit the deformation due to traffic load repetitions by furnishing homogeneous stress distribution layer. Asphalt stabilization technique is usually used to control and improve poor subgrade soil condition. In this work, the unconsolidated undrained Triaxial test was implemented for comparatively assessing the impact of asphalt stabilization on shear strength and stiffness of asphalt stabilized subgrade soil. Soil samples have been treated with cutback asphalt using various water and liquid asphalt percentages. The loose asphalt stabilized soil samples were subjected to aeration periods ranging from one to five hours at room temperature of 20 ± 2 °C before compaction. Cylindrical Specimens with height and diameter of (77.4 and 38) mm respectively have been prepared in the laboratory after aeration periods using static compaction to achieve a target density. The compacted Specimens have been subjected to curing at room temperature of 20 ± 2 °C for a curing period ranging from seven to ninety days, then tested in the Triaxial apparatus to determine the shear strength and modulus of elasticity properties. Data were analyzed and compared. It was concluded that the optimum aeration period was two hours while the reasonable curing period is seven days. It was observed that the shear strength and the modulus of elasticity increases by (211 and 251) % respectively with increasing cutback asphalt percentage up to 8% asphalt content, as compared with natural soil. It was concluded that the effect of variation in fluid content in the range of 0.5% above or below the optimum has increased the shear strength by (663, and 662) % respectively as compared with natural subgrade soil.

Keywords: Liquid asphalt · Subgrade · Shear strength · Aeration · Curing · Triaxial test

1 Introduction

Subgrade materials should be characterized by their engineering properties such as strength and stiffness (stress-strain relationship under traffic loading) in order to control pavement damage. Providing efficient stiffness for subgrade cannot be easily obtained all the time, (Sarsam et al. 2013); therefore, improving the engineering properties of subgrade soils using stabilizations could provide a suitable solution. MC-30 bituminous

© Springer Nature Switzerland AG 2020
H. Shehata et al. (Eds.): GeoMEast 2019, SUCI, pp. 70–80, 2020.
https://doi.org/10.1007/978-3-030-34184-8_4

cutback has been used by (Kumar and Bansal 2017) as a stabilizer to improve the properties of cohessionless soil. The percentage of bituminous cutback added to the sandy soil has been varied from 4% to 18%. Unconfined Compressive Strength and CBR value of the soil samples were studied. It was concluded that maximum unconfined compressive strength is obtained at 12% cutback content, further increase of cutback in the soil leads to decrease in UCS value due to excessive fluidity causing decrease in density. A sharp increase in CBR value of sand was observed with a small increment in cutback bitumen content (about 8%). The deformation behavior of emulsified asphalt stabilized embankment model under monotonic and cyclic loading was investigated by (Sarsam et al. 2014). It was concluded that the ultimate sustained pressure was 0.8 MPa with vertical settlement 0.03 mm for pure soil at dry condition while it was reduced to 0.3 MPa with vertical settlement 12 mm at absorbed condition. It was concluded that addition of asphalt emulsion has positive impact on soil behavior; the embankment model was able to sustain an applied stress of (4–4.5 MPa) before failure and shows 0.12 mm of vertical deformation at dry and absorbed test conditions under monotonic loading. Under cyclic load condition, the stabilized soil was able to sustain 911 and 897 load repetitions at failure for dry and absorbed test conditions respectively. Soil stabilization using emulsified asphalt as stabilization material has been studied by (Bunga et al. 2011). The results indicate that emulsified asphalt can improve physical, chemical, and mechanical characteristics of sandy clay loam. Plasticity and shear strength of soil increase with the increase of emulsified asphalt concentration. Chemical binding occurs between minerals in the soil and chemical elements by using emulsified asphalt. As explained by (Sarsam and Ibrahim 2008) the addition of cutback Asphalt to soil has a negative impact on cohesion, while the angle of internal friction increases. Lower cutback content of 5% shows higher shear strength at high normal stresses. This may indicate that lower cutback content could give higher shear strength. It was concluded that Cementing and waterproofing qualities of soil can be improved by using asphalt as a stabilizing agent. The cementation property is most effective in providing increased stability of soil. A series of laboratory tests have been conducted by (Avinash et al. 2015) namely, USCS soil classification, specific gravity, optimum moisture content, maximum dry density, liquid limit, plastic limit, California bearing ratio are conducted on soil samples with varied bitumen content ranging by (0, 1, 5, 9, and 13) %, and tested thereafter. The results stand out the increase in liquid and plastic limits on addition of optimum cutback asphalt. However, in case of stability there is an increase in California bearing ratio in soaked and dry condition with increase in cutback.

The aim of this investigation is to study the effect of asphalt content, aeration and curing periods on shear strength and modulus of elasticity of cutback asphalt stabilized soil by implementation of unconsolidated undrained Triaxial test.

2 Materials and Methods

2.1 The Subgrade Soil

The sub-grade soil was obtained from Al-Taji city, north of Baghdad, the soil was excavated from a depth of (1 to 2.5 m) after removal of the top soil. Grain Size distribution of this Soil was found by Sieve analysis. The results are shown in Fig. 1. Soil is classified as (SM) by Unified Soil Classification System (USCS) according to (ASTM D 2487 2009) Using American Association of State Highway and Transportation Officials (AASHTO 2013). The subgrade soil was classified as (A-1b). Table 1 shows the chemical composition of the soil, while Table 2 presents the physical and geotechnical properties of the soil.

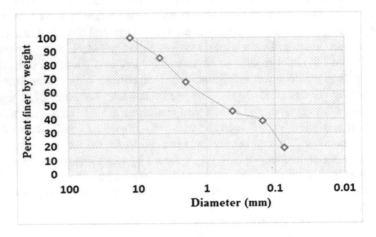

Fig. 1. Grain size distribution of the soil

Table 1. Chemical composition of the soil

Chemical composition	Test result %
Total (SO_3)	0.712%
Carbonate content (Ca Co_3)	1.069%
Calcium sulfate (Ca SO_4)	1.111%
Total soluble salts (T.S.S.)	1.31%
PH value	10.03

Table 2. Physical and geotechnical properties of the soil

Property	Test results
Percent passing 0.075 mm sieve	18.7
AASHTO classification (Sulaiman and Sarsam 2000)	A-1b
Unified soil classification	SM
Specific gravity	2.64
Liquid limit %	23
Plasticity index %	Non-plastic
Maximum dry density (gm/cm^3) (modified compaction)	1.76
Optimum water content %	16
Cohesion kPa, (direct shear box)	41
Angle of internal friction	29.2
Undrained shear strength kPa (unconfined compression test)	50
Unconsolidated undrained shear strength kPa (triaxial test, 100 kPa cell pressure)	54

2.2   Cutback Asphalt

The type of liquid Asphalt used in this study was Medium curing Cutback Asphalt (MC-30) that had been produced according to (ASTM D 2027 2009), and (AASHTO - M 82 2013), by Al - Dora Refinery using one step, it is composed of 91.2% asphalt cement of grade 85–100, and 8.8% Kerosene. The Properties of Cutback Asphalt (MC-30) as supplied by the refinery are illustrated in Table 3. This grade of cutback gives higher dry density than other grades of Cutback Asphalt due to less Viscosity of (MC-30) and has more solvent content which causes better mixing and coating of Soil's Particles and better compaction.

Table 3. Properties of cutback asphalt (MC - 30) as supplied by Dora refinery

Property	Results
Flash point (C.O.C) °C (min.)	38
Viscosity (C. st.) @ 60 °C	30–60
Water % V (max.)	0.2
Distillation test to 360 °C, distillate % V of total distilled	
To 225 °C (max.)	25
To 260 °C (max.)	40–70
To 315 °C (max.)	75–93
Residue from distillation to 360 °C % V (min)	50
Tests on residue from distillation	
Penetration @ 25 °C (100 g, 5 s, 0.1 mm)	120–250
Ductility @ 25 °C (cm) (min)	100
Solubility in tri-chloro ethylene % wt. (min)	99

2.3   Preparation and Testing of Specimens

The dry soil was mixed with optimum water content for two minutes to become homogenous then it was mixed with the required percentage of cutback asphalt for three minutes so that the soil particles are covered with thin film of asphalt. The mixture was left for aeration at room temperature of 25 °C for various periods ranging from one to five hours before compaction. Figure 2 shows mixture under aeration process. The procedure of obtaining the optimum percent of cutback asphalt is published elsewhere, (Sarsam and AL Sandok 2017) and (AL Sandok 2018). The mixture was then transferred into the mold and subjected to static compaction to a target dry density of 1.760 gm/cm^3. Specimens were prepared with optimum fluid content (water + cutback asphalt) of 16%. Specimens were left for curing at room temperature of 25 ± 2 °C for various periods ranging from seven to ninety days before testing. A total of 15 Stabilized soil specimens of 77.4 mm height and 38 mm diameter have been prepared in the laboratory after aeration periods using static compaction to achieve a target density. Then Another 15 Specimens of 77.4 mm height and 38 mm diameter have been prepared in the laboratory after the optimum aeration period using static compaction and subjected to various curing periods as mentioned above and tested for shear properties in the Triaxial apparatuses.

Fig. 2. Aeration process of asphalt stabilized soil

2.4   Determination of the Optimum Fluid Content using Unconsolidated Undrained Triaxial Test

This test was carried out on specimens of pure (natural) and stabilized soil with optimum water content and (4%, 6%, 8% and 10%) of MC-30 cutback liquid asphalt implemented as partial replacement of water content. All the specimens were left for two hours for aeration before compaction, and then for seven days to cure before testing. The objective of these aeration and curing processes is to remove the excess fluid content (water and kerosene) to furnish a proper compaction to achieve the

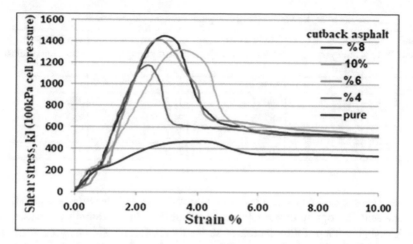

Fig. 3. Stress-strain relationship for stabilized and natural soil under triaxial test

maximum shear strength and maximum modulus of elasticity. The stress-strain relationship of stabilized and natural soil is shown in Fig. 3. The shear stress increases until it reaches a peak value then it decreases as the strain increases. The variation of shear strength was not significant at early stage of loading up to 1% strain, then the shear strength was variable for different asphalt percentages as compared to natural soil.

3 Results and Discussion

It can be observed from the Fig. 4 and Table 4 that the shear strength obtained by Triaxial test increases with increasing cutback asphalt percentage up to 8% asphalt content, then it decreases dramatically. This amelioration may be brought about due to an increased cohesion potential, which is caused by a coating of asphalt film to the soil

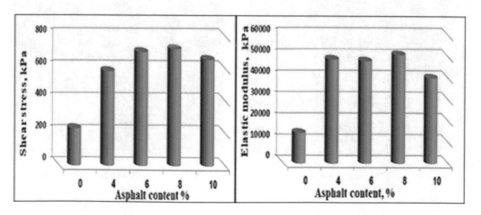

Fig. 4. Variation of shear strength and elastic modulus with asphalt content.

Table 4. Shear strength-asphalt content (%) relationship

Asphalt content (%)	Shear strength (kPa)	Modulus of elasticity (kPa)	% Change in shear strength	% Change in modulus of elasticity
Natural	233	14588	–	–
4	588	49057	152	236
6	707	48233	203	231
8	726	51249	211	251
10	661	40397	183	177

particles. The shear strength reaches a maximum value at 8% cutback asphalt, which may represent the optimum particle coating, but the shear strength decreases as the cutback asphalt content future increases above the optimum. This may be attributed to the increases in thickness of asphalt films surrounding the soil particles and the fluid content is filling more voids, preventing the particle interlock, which causes a high reduction in friction, and which in turn, leads to a reduction in the shear strength. These results appear to be in accordance with several researchers' findings, (Al-Safarani 2007); (Sarsam and Barakhas 2015); (Sulaiman and Sarsam 2000) and (Taha et al. 2008). It can be noted the modulus of elasticity reaches its peak value at 8% asphalt content then it decreases with further increment in asphalt content. The addition of the optimum fluid content of 16% (8% cutback asphalt + 8% water) has increased the shear strength and the modulus of elasticity (E), by 211% and 251% respectively, as compared with pure subgrade soil condition. Figure 5 exhibits the test setup of the specimens, while Fig. 6 shows part of the prepared specimens under curing.

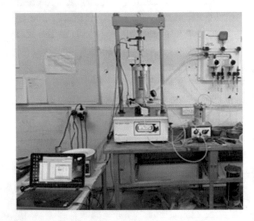

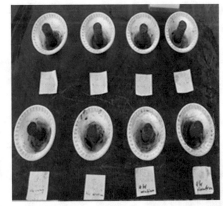

Fig. 5. Test setup of specimen

Fig. 6. Part of the prepared specimens

3.1 Influence of Aeration Period

Figure 7 exhibit the stress strain relationship of asphalt stabilized soil after it was subjected to various aeration periods. The highest shear stress could be sustained by the specimen after two hours of aeration. On the other hand, five hours of aeration exhibit the lowest shear stress than the specimen could practice. Figure 8 summarizes the variation in shear strength of asphalt stabilized soil specimens among various aeration periods. It can be noted that two hours of aeration is capable to provide the optimum aeration period while the shear strength decreases as the aeration period increases. This could be attributed to the fact that an optimum fluid content is required during compaction to achieve the required density. Figure 9 exhibit the impact of aeration period on the stiffness of asphalt stabilized soil, it can be noted that two hours of aeration exhibit the optimum time to gain the highest stiffness among other aeration periods.

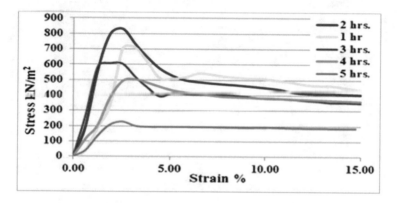

Fig. 7. Stress-strain relationship at various aeration periods

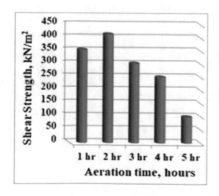

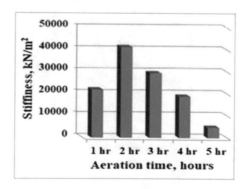

Fig. 8. Variation in shear after aeration   **Fig. 9.** Variation in stiffness after aeration

3.2   Influence of Curing Period

Figure 10 exhibit the stress strain relationship of asphalt stabilized soil after it was subjected to various curing periods. The highest shear stress could be sustained by the specimen after ninety days of curing at laboratory environment. On the other hand, seven days of curing exhibit the suitable shear stress that the specimen could practice. As demonstrated in Fig. 11, the shear strength increases sharply at early ages as compared to that after 21 days, while a reasonable shear strength could be obtained after seven days of curing. Such finding agrees well with (Sarsam and Ibrahim 2008). On the other hand, Fig. 12 exhibit the variation in the stiffness of asphalt stabilized soil among various curing periods. It can be observed that the variation is not significant at lower curing periods.

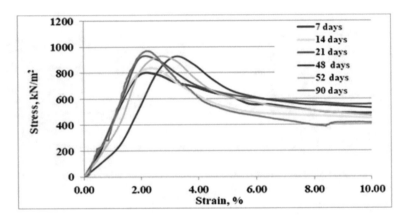

Fig. 10. Stress-strain relationship at various curing periods

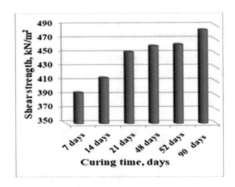

Fig. 11. Variation in shear after curing

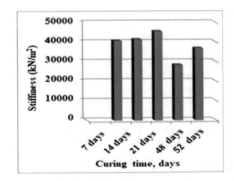

Fig. 12. Variation in stiffness after curing

3.3   Influence of Variable Asphalt Content

Figure 13 shows the shear stress of the Unconsolidated Undrained Triaxial Test with cell pressure of 100 kPa for stabilized with several percentages of fluid content and natural soil. The addition of (optimum-0.5, optimum, optimum+0.5) % of fluid content has increased the shear strength by (663, 666, 662) % respectively, as compared with natural (pure) subgrade soil, while the effect of variation in fluid content in the range of 0.5% above or below the optimum fluid content was not significant as demonstrated in Fig. 14.

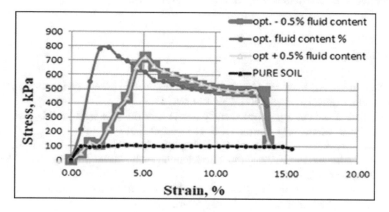

Fig. 13. Stress-strain relationship for stabilized and natural soil.

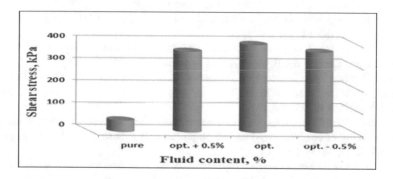

Fig. 14. Shear strength for asphalt stabilized and natural soil

4   Conclusions

Based on the testing program, the following conclusions may be drawn:

1. The shear strength and the modulus of elasticity increases by (211 and 251) % respectively with increasing cutback asphalt percentage up to 8% asphalt content, then it decreases dramatically as compared with natural soil.

2. Two hours of aeration is capable to provide the optimum aeration period while the shear strength and the stiffness decreases as the aeration period increases.
3. The highest shear stress could be sustained by the specimen after ninety days of curing at laboratory environment. On the other hand, seven days of curing exhibit the reasonable shear stress that the specimen could practice. The variation in the stiffness of asphalt stabilized soil among various curing periods is not significant at lower curing periods.
4. The effect of variation in fluid content in the range of 0.5% above or below the optimum fluid content has increased the shear strength by (663, and 662) % respectively as compared with natural subgrade soil.

References

Sarsam, S., Alsaidi, A., Mukhlef, O.: Behavior of reinforced gypseous soil embankment model under cyclic loading. J. Eng. **19**(7), 830–844 (2013)
Kumar, V., Bansal, R.: An experimental study on the behavior of a sandy soil by using cutback bitumen. Int. J. Multidiscip. Curr. Res. **5**, 1134–1137 (2017)
Sarsam, S.I., Alsaidi, A.A., Alzobaie, O.M.: Impact of asphalt stabilization on deformation behavior of reinforced soil embankment model under cyclic loading. J. Eng. Geol. Hydrogeol. JEGH **2**(4), 46–53 (2014)
Bunga, E., Pallu, S., Selintung, M., Thaha, A.: Stabilization of sandy clay loam with emulsified asphalt. Int. J. Civil Environ. Eng. **11**(05), 52–62 (2011)
Sarsam, S., Ibrahim, S.: Contribution of liquid Asphalt in shear strength and rebound consolidation behavior of Gypseous soil. Eng. Technol. **26**(4), 484–495 (2008)
Avinash, P., Praneeth, S., Sekhar, P., Sahu, M., Siddhardha, R., Chandra, S.: Stabilization of soil in the capital region of Andhra Pradesh using cutback asphalt. HCTL Open Int. J. Technol. Innov. Res. (IJTIR) **14**, 1–8 (2015)
American Society for Testing and Materials, ASTM: Road and Paving Material, Vehicle-Pavement System, Annual Book of ASTM Standards, vol. 04.03 (2009)
AASHTO: Standard Specification for Transportation Materials and Methods of Sampling and Testing, American Association of State Highway and Transportation Officials, 14th edn, Part II, Washington, D.C. (2013)
Sarsam, S., AL Sandok, A.: Comparative assessment of deformation under repeated loading and wheel tracking of reinforced and asphalt stabilized subgrade soil. Int. J. Emerg. Eng. Res. Technol. **5**(12), 38–44 (2017)
Al Sandok, A.: Verification of layered theory for stabilized and reinforced subgrade model. M. Sc. thesis, Department of Civil Engineering, University of Baghdad, Iraq (2018)
Al-Safarani, M.G.: Improvement of gypseous soil characteristics using cutback asphalt and lime. M.Sc. thesis, Civil Engineering Department, University of Al – Mustansiria (2007)
Sarsam, S., Barakhas, S.: Assessing the structural properties of asphalt stabilized subgrade soil. Int. J. Sci. Res. Knowl. (IJSRK) **3**(9), 0227-240 (2015)
Sulaiman, R.M., Sarsam, S.: Effect of liquid Asphalt on geotechnical properties of Gypseous soil. Eng. Technol. **19**(4), 1–18 (2000)
Taha, M., Al-Obaydi, A., Taha, O.: The use of liquid asphalt to improve gypseous soils. AL-Rafidain Eng. **16**(4), 38–48 (2008)

Motorway Rockfill Embankments

Anastasios Mouratidis[1] and Apostolis Ritsos[2(✉)]

[1] School of Civil Engineering, Aristotle University of Thessaloniki AUTH,
Thessaloniki, Greece
anmourat@civil.auth.gr
[2] Geotechnical Civil Engineer NTUA, Edafomichaniki S.A., 19 Emm. Papadaki,
14121 Athens, Greece
aritsos@edafomichaniki.gr

Abstract. The construction of Rockfill Embankments, although considered as a process of low difficulty, it presupposes the compliance of the design study with rigorous specifications. In this regard, a meticulous and systematic method of construction is required to ensure the optimum performance of this type of geostructure. In this paper, some basic requirements concerning the design and construction of rockfill embankments, as well as some engineering recommendations for a successful construction, are presented and commented. Advantages and weaknesses of rockfill embankment, in comparison to the traditional earthfill structures are also illustrated and the relevant fields of application are defined. Finally, a case study, dealing with the construction of a high motorway geostructure, consisting of crushed rock material at its lowest part and of reinforced earthfill at the top, is briefly presented.

1 Introduction

Rockfill Embankments are structures composed of natural, durable rock materials, free-drained, where coarse grained materials, such as gravel, cobbles, boulders, are predominant. Rockfill may also include, in total or partly, small amount of sandy and fine-grained materials, according to a specific gradation. On site, rockfill material is spread in horizontal layers, compacted by heavy vibratory rollers. Sound rock is the most suitable material but some weathered or weak rocks may also be utilized, including sandstones and cemented shales, under rigorous conditions. Fragile rocks, susceptible to break down to fine particles during excavation, should be considered and treated as earthfill. In order to meet construction requirements, excess fine material and inadequate rock fragments, should be removed, while oversized stones should break into smaller particles.

2 Classification of Borrow and Excavation Material

Physical and strength testing of materials is a prerequisite in embankment construction. In rockfill construction, this includes the crushed rock, the coarse gravel and the fine soil. All available materials are classified into soil and rock material groups. Identification involves the characterization of soils and rocks by laboratory testing after the

© Springer Nature Switzerland AG 2020
H. Shehata et al. (Eds.): GeoMEast 2019, SUCI, pp. 81–93, 2020.
https://doi.org/10.1007/978-3-030-34184-8_5

excavation stage. This usually includes gradation, water content, carbonates, sulphates, organic content, plasticity and liquidity limits for fine material, stone size, stone density, specific gravity and unit weight. Laboratory strength tests may comprise shear strength along discontinuities, uniaxial compressive strength (UCS), tensile strength - Brazilian Test, Point Load test index (I_{PLT}), fragmentability, friability, degradability, Los Angeles index, micro Deval index, frost resistance for rocks and specific tests that are usually specified at the relative geotechnical design.

The basic classification of the soil and rock groups are summarised in the following tables: Table 5 refers to the rock material groups with regard to their strength (prEN 16907-2, table 4a). Tables 6-1 & 6-2 refers to the particle size fractions (according to ISO/FDIS 14689-1 table A1, 14688-1 table 1, 14688-2 table 1, prEN 16907-2 table 3, [7]). Tables 7-1 and 7-2 refers to the classification of rock material groups for reuse after excavation and fraction (prEN 16907-2, table 4b).

3 Motorway Design of Rockfill Embankments

Embankments are the principal geotechnical structures in road engineering. They may be homogenous or divided into different zones (prEN 16907-1, Figs. 1 and 2) in relation with materials available: (A) base-foundation, (B) core, (C) berms-shoulders (D) transition layers, (L) capping layers, (S) pavement superstructure. A road embankment is defined by its geometry, that is, its height "H", the slope angle "I", the spacing of berms h = 7–12 m, the width of berms b = 4–5 m. The slope angle usually varies: in rockfill from 2:3 to 1:1 (34° to 45°, height: base), and in earthfill from 1:2 to 2:3 (27° to 34°, height: base).

Fig. 1. Construction of an embankment with rockfill material

Fig. 2. Crushed rock materials used for the construction of a road and a railway alignment

Rockfill material specifications for embankment construction are characterised by its gradation. The percentage of material passing through sieve 1' or 3/4" (ASTM) must be less than 30%. Alternatively, material greater than 4 inches must exceed 25% (USA). However, these conditions are not sufficient to prescribe the suitability of construction material. Rockfill can contain intermediate and fine particles in order to form a resistant structure, hardly deformable and definitely durable. This leads to additional requirements for the rockfill material.

A satisfactory gradation, includes a continuous grain curve, and the passing material percentages (%) for different sieve sizes should comply with the following:

Sieve size	D	D/4	D/16	D/64
Passing %	90-100	45-60	25-45	15-35

When gradation and soundness requirements are met, rockfill construction proves economical because the excavated material can be directly used without processing. Rock fill material, due to its high shear strength, can be used in motorway embankments when difficult morphological conditions are encountered, especially when the ground water level is high or the ground profile is steep. Moreover, due to its low deformability and susceptibility to self-settlement, rockfill material can be used in the construction of very high embankments (H > 30 m), and, in case of adequate foundation conditions, for the replacement of a viaduct structure.

The most important feature of a rockfill material for road embankments is its high shear strength. Relatively, the earthfill materials usually have an effective angle of internal friction varying from 25° to 35°, while the rockfill materials usually have an effective angle of internal friction varying from 35° to 50°. A design value of $\varphi' = 35°$ usually is conservatively proposed for compacted granular material. After conducting shear box tests on well graded granular materials, compacted to the required density, the design friction angle was calculated to: $\varphi'_{PEAK} = 40°–45°$ [6]. An average friction angle of around 50° can be expected for well compacted, good quality, clean, dumped rock material [8].

For common practice, the conservatively proposed design values concerning the mechanical properties of an embankment fill, are summarised in Table 1.

Table 1. Proposed design values for compacted coarse to very coarse grained fill

	φ' (°)	c' (kPa)	γ (kN/m³)	Es (MPa)
Rockfill	36–40 (≤ 45)	0–2	21–23	≥ 100
Crushed rock to sand-gravel	34–38	3–5	20–21	70–150
Coarse sand	30–34	3–5	19–21	50–100
Fine sand	27–30	3–5	19–21	40–80
Sanitary layer	30–36	0	20–21	30–50

φ': effective angle of internal friction
c': effective cohesion
γ: material wet bulk density
Es: fill mass modulus of compressibility

Insignificant embankment deformation, under permanent and traffic loads, is a major advantage of rockfill embankments. Field measurements of subsidence range from 0,5‰ H to 1‰ H for rockfill embankments and from 0,5% H to 1% H for earthfill embankments.

Geosynthetic industrial materials, such as geogrids (mesh opening equal to the diameter of the 50% passing D50 of the fill material), geotextiles, geonets, geocomposites, etc. can be used for basal and/or body reinforcement, in order to increase the fill materials shear strength, for filtration and for soil separation, in order to prevent migration of the finer material into the voids of the coarser layers, especially where the ground water is encountered. Geosynthetic industrial materials are usually mandatory in the base of rockfill embankments, where large stones or cobbles are encountered as the constitutive material over fine natural ground material.

4　Rockfill Construction Principles and Practices

The basis of the fill construction [2] is about "defining the process" in order to construct a well compacted and durable embankment, after the characterization of the natural ground, the selection of the suitable equipment and setting the rules to design the extraction, the transportation, the compaction and the control of the chosen fill materials.

Dozers and vibrating rollers are the most suitable construction equipment in construction of rockfill embankments. Large pieces of degradable material should be broken down before spreading the material on a layer. Alternatively, padfoot or sheepsfoot rollers may be used to crush large boulders. Layers of 50–100 cm thickness make the rule in rockfill embankment construction, whereas earthfill layers can hardly exceed a thickness of 50 cm (Tables 2 and 3). In any case the maximum size of the rockfill material must not exceed 2/3 of the lift layer thickness.

Table 2. Typical ranges: Bulking Factor (BF) - Compaction Factor (CF) - Lift layers thickness after compaction - % Compaction

Soil/rock	Clay	Sand	Sandy gravel	Gravel	Siltstone claystone	Limestone sandstone	Soft rock	Crushed rock
BF after loosening	0,75 0,85	0,80 0,90	0,80 0,85	0,75 0,80	0,75 0,80	0,65 0,75	0,60 0,70	0,20 0,70
CF after compaction	1,00 1,10	1,05 1,20	1,05 1,20	0,90 1,00	0,85 1,00	0,75 0,90	0,70 0,80	0,30 0,80
Lift layer thickness (mm)	200 400	300 700	400 800	500 900	300 700	500 1000	500 800	500 1500
Compaction (%)	96 100	92 98	92 96	94 96	Survey	Survey	Survey	Survey

Table 3. Rockfill & Earthfill - recommended construction practices

Performance criteria	Earthfill		Rockfill	
	Fine soil	Coarse soil	Degradable	Sound
Gradation, soundness	>35% passing from sieve No200 (0,074 mm)	<35% passing from sieve No200 (0,074 mm)	soundness (Na$_2$SO$_4$) > 10%	soundness (Na$_2$SO$_4$) < 10%
AASHTO, ASTM D3282 classification	A-4, A-5, A-6	A-1, A-2-4, A-2-5	siltstone, schist, (Tables 6-1, 6 2)	granite, gneiss, limestone (Tables 6-1, 6-2 and 7-1, 7-2)
EN classification	cSi, mSi, fSi, Cl	cGr, mGr, fGr, cSa, mSa, fSa	(Tables 7-1, 7-2)	(Tables 7-1, 7-2)
Usual embankment max height	H < 15 m	H < 30 m	H < 30 m	H < 50 m
Recommended lift layers (mm)	200–300	300–500	500–1000	500–1000
Construction equipment	Graders	Graders, dozers	Dozers	Dozers

(*continued*)

Table 3. (*continued*)

Performance criteria	Earthfill		Rockfill	
	Fine soil	Coarse soil	Degradable	Sound
Roller type	Sheepsfoot, padfoot	Vibrating	Sheepsfoot-vibrating	Vibrating
Compaction control	Sand & cone method, nuclear by troxler	Sand & cone method, nuclear by troxler	Steel slab, CCC, plate load test	CCC, steel slab, plate load test
Main use of geosynthetics	Core reinforcement	Basal, core reinforcement	Basal, core reinforcement	Basal reinforcement
Embankment deformation	0,50–1,0% H	0,25–0,50% H	1‰ H	0,5‰ H
Risk of slope failure	Significant	Intermediate	Intermediate	Negligible
Risk of slope erosion	Intermediate	Significant	Significant	Intermediate
Shoulders water-proof	Mandatory	Useful	Mandatory	Useful
Overall assessment	Fair	Good	Good	Excellent

Table 4. Case Study - Geotechnical design parameters

	φ' (°)	c' (kPa)	γ (kN/m³)	Es (MPa)	Group symbol	Geotechnical parameters (σ_{ci} UCS I_{PLT}, units in MPa)
Earthfill	35	0–5	20	50	MSa cSa mGr cGr	lift layers 600 mm, sand-gravel, IP < 10%
Rockfill borrow materials from cuts excavations	45	0	23	≥ 100	RMS-RS	lift layers 750 mm L.A. < 40% aggregates soundness < 8%, dmax = 50 cm, maxW200kgr, 25% < 5kgr
Sanitary layer 30 cm thick	35	0	20	50	CSa cGr	Coarse grained material, fines < 7%
Schist-chert (sch)	30–37	90	25,5	800	RMS-RS	GSI = 32–48 mi = 15 σ_{ci} = 27 UCS = 32–69 I_{PLT} = 1,7
Serpentinite (se)	27	70	25	150	RW-RMS	GSI = 22–38 mi = 13 σ_{ci} = 11 UCS = 15 I_{PLT} = 1,3

Table 5. Classification of Rock material groups, related to their strength (prEN 16907-2, table 4a)

Rock group		Point Load Index I_{s50} Assuming $I_{s50} = \sigma_u/25$ (MPa)	Compressive strength σ_u (MPa)	Rock examples
Rock group symbol	Strength term			
REW	Extremely weak rock		0,6–1	Weathered claystone, siltstone, sandstone
RVW	Very weak rock	Less than <0,2	1–5	Weathered claystone, siltstone, sandstone, gypsum-stone, coal
RW	Weak rock	0,2–1,0	5–25	Claystone, siltstone, sandstone, marlstone, schists, gypsum-stone, coal
RMS	Medium strong rock	1,0–2,0	25–50	Sandstone, marlstone, limestone, schists, metamorphic rock
RS	Strong rock	2,0–4,0	50–100	Sandstone, limestone, volcanic rock, plutonic rock, metamorphic rock
RVS	Very strong rock	4,0–10,0	100–250	Volcanic rock, plutonic rock, metamorphic rock
RES	Extremely strong rock	>10,0	>250	

NOTE 1: RVW and RW materials may be degradable and testing may be necessary to confirm the material stability
NOTE 2: Alternative material specific correlation of I_{s50} with σ_u may be used when available
NOTE 3: Other intrinsic properties of rock used in classification may include mineralogy and density

In most cases, no additional water is needed to achieve proper compaction. Vibratory rollers and sheepsfoot rollers are used in the compaction of rockfill material. Rollers must have operating weight from 12–25 ton, capable of compacting coarse material in lift layers of up to 1500 mm. The required compaction is first checked over a trial section and the best result can usually be obtained after 4–8 passes of the roller.

Bulking factor "BF" is the ratio of borrow material volume, prior and after loosening. Compaction factor "CF" is the ratio of borrow material volume, prior to loosening and after compaction. The proposed typical range of those parameters are given in Table 2.

Drainage conditions during construction and lifetime are critical, mainly for stability purposes. Waterproofing and drainage measures are essential to the integrity of

Table 6-1. Particle size fractions (according to ISO/FDIS 14689-1 table A1, 14688-1 table 1, 14688-2 table 1, prEN 16907-2 table 3, [7])

Grain size	Particle size fractions (symbol)	Range of particle sizes (mm)	Primary fraction	Rock fragments (fractions are depended to excavation methods)
Very coarse soil group 1 (fractions less than 63 mm: $\leq 30\%$) soil group 2 (fractions less than 63 mm: $\leq 70\%$)	Large Boulder (lBo)	>630		Sedimentary: conglomerate (rounded fragments), Breccia (angular fragments), Halite gypsum Pyroclastic: Agglomerate (round grains), volcanic breccia (angular grains) Igneous (quartz content): granite, aplite, granodiorite, diorite, gabbro, peridotite Metamorphic: gneiss, migmatite, marble, quartzite, granulite, hornfels
	Boulder (Bo)	>200 and ≤ 630	>50% of particles by mass ≥ 200 mm	
	Cobble (Co)	>63 and ≤ 200	>50% of particles by mass <200 mm and ≥ 63 mm	

Table 6-2. Particle size fractions (according to ISO/FDIS 14689-1 table A1, 14688-1 table 1, 14688-2 table 1, prEN 16907-2 table 3, [7])

Grain size	Particle size fractions (symbol)	Range of particle sizes (mm)	Primary fraction	Rock fragments (fractions are depended to excavation methods)
Coarse soil (fine less than 0,063 mm: <5%) Composite coarse soil (fine less than 0,063 mm: 5% to 15%)	Gravel (Gr)	>2,0 and ≤ 63	>50% of particles by mass < 63 mm and ≥ 2 mm	
	Coarse gravel, cGr	>20 and ≤ 63		
	Medium gravel, mGr	>6,3 and ≤ 20		
	Fine gravel, fGr	>2,0 and $\leq 6,3$		

(*continued*)

Table 6-2. (*continued*)

Grain size	Particle size fractions (symbol)	Range of particle sizes (mm)	Primary fraction	Rock fragments (fractions are depended to excavation methods)
	Sand (Sa)	>0,063 and ≤2,0	>50% of particles by mass < 2 mm and ≥0,063 mm	Sedimentary: sandstone, quartzite, arkose, greywacke, chalk, lignite, coal Pyroclastic: coarse grained tuff Igneous: microgranite, microdiorite, dolerite Metamorphic: schist, serpentine
	Coarse sand (cSa)	>0,63 and ≤2,0		
	Medium sand (mSa)	>0,20 and ≤0,63		
	Fine sand (fSa)	>0,063 and ≤0,20		
Fine soil (fine less than 0,063 mm: >35%) Intermediate soil (fine less than 0,063 mm: 15% to ≤35%)	Silt (Si)	>0,002 and ≤0,063	low plasticity or non-plastic	Sedimentary: mudstone, siltstone Pyroclastic: fine grained tuff Igneous: rhyolite, dacite, andesite, quartz, trachyte, basalt Metamorphic: phyllite, slate
	Coarse silt (cSi)	>0,02 and ≤0,063		
	Medium silt (mSi)	>0,0063 and ≤0,02		
	Fine silt (fSi)	>0,002 and ≤0,0063		
	Clay (Cl)	≤0,002	plastic	Sedimentary: shale, claystone Pyroclastic: very fine-grained tuff Igneous: rhyolite, dacite, andesite, quartz, trachyte, basalt Metamorphic: phyllite, slate

Table 7-1. Classification of Rock material groups for reuse after excavation (prEN 16907-2, table 4b)

Indicative strength	Rock groups for reuse	Parameters (Intrinsic properties)						Symbol	Behaviour	Comments
	Geological nature	Compressive strength (MPa)	FragmentabilityFR	DegradabilityDG	Los Angeles Index LA	Micro deval Index MDE	Density Yd			
Weaker	Clay rocks	<5	>7					CRw	As a soil after extraction	Usable in earth structure
	Limestone					>45	<1,8	LIw		
	Sandstone		>7					SAw		
	Conglomerate		>7					COw		
	Igneous rocks		>7					IRw		
	Metamorphic rocks		>7					MRw		
Intermediate	Clay rocks	5–50	>7	>5 >2				CRid	Evolutive or degradable rock	Depending on the design
	Clay rocks		<7	<5 <2				CRind	Non evolutive or non-degradable rock	Usable in earth structure
	Limestone					>45	>1,8	LIi	Depending on the earth work procedure	
	Sandstone		<7		>45	>45		SAi		
	Conglomerate		<7		>45	>45		COi		
	Igneous rocks		<7		>45	>45		IRi		
	Metamorphic rocks		<7		>45	>45		MRi		
Strong	Clay rocks	50–100	<7	>5 >2		<45		CRSd	Evolutive or degradable rock	Depending on the design
	Clay rocks		<7	<5 <2		<45		CRSnd	Non evolutive or non-degradable rock	Usable in earth structure
	Limestone		<7	>5 >2		<45	no limit	LIS	As a granular soil	
	Sandstone		<7	<5 <2	<45	<45		SAS		
	Conglomerate				<45	<45		COS		
	Igneous rocks				<45	<45		IRS		
	Metamorphic rocks				<45	<45		MRS		

Table 7-2. Classification of Rock material groups for reuse after excavation (prEN 16907-2, table 4b)

Indicative strength	Rock groups for reuse									
	Geological nature	Parameters (Intrinsic properties)						Symbol	Behaviour	Comments
		Compressive strength (MPa)	Fragmentability FR	Degradability DG	Los Angeles Index LA	Micro deval Index MDE	Density Yd			
Very strong	Igneous and metamorphic rocks	100–250			<35	<25		IRVSMRVS		
Extremely strong	Igneous and metamorphic rocks	>250			<25	<10		IRESMRES		

NOTE: The values for FR and DG will vary depending on the test procedure adopted. In this Table: parameters for FR according to French standard NF P 94-06; parameters for DG according to French standard NF P 94-067 (first value) and Spanish standard UNE 146510 (second value)

the geotechnical structure. Rockfill embankments are self-draining, while for draining purpose, the excavation level should have a transverse grade of 1% to 4%.

In rockfill embankments higher than H > 15 m, where applied loads are significant, fragmentation may occur due to high super-incumbent pressure. In order to prevent high embankment deformation, rock fragments RMS, RS, RVS, RES (Table 5) with compressive strength $\geq$ 25 MPa should be used in construction.

Concerning field quality control, in rock fills containing boulders of particle size greater than 200 mm, the estimation of the compaction degree using the traditional sand and cone method is difficult or impossible. Field quality control usually consists of levelling measurements and steel slab bearing tests. The levelling tests consist of measuring the settlement of the compacted lift (usually by means of a steel slab, with an appropriate thickness, placed on the compacted surface) after one pass of the dozer or after as many passes as to obliterate vertical displacement [2, 9]. In several cases, continuous compaction control (CCC) by specific equipment of vibrating rollers is conducted.

For rockfill material, containing at least 6% of fines, the most common field tests are: sampling and suitability laboratory tests, in-situ density (nuclear method by a Troxler equipment, sand replacement, etc.), plate load tests (static, dynamic).

Monitoring of the embankment's fill deformations, mainly during the construction, is performed using settlement plates, settlement cells, slope inclinometers, piezometers.

Trial sections must precede construction to check material suitability and effectiveness of spreading and compaction equipment.

5 Case Study – Reinforced Earth Embankment on Rockfill Embankment

A motorway embankment of a 500 m section was constructed next to a lake, having maximum height of H $\leq$ 45 m. Its lower part consists of a rockfill, 15 m high, which is tangential to the reservoir. Its upper part consists of a reinforced earthfill, 30 m high. The fill was founded on the Alpine formations (Triassic - Jurassic) of Schist-chert (sch) and on the ophiolitic schistose serpentinite (se), rock formations that have high

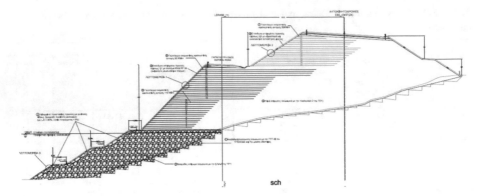

Fig. 3. Case study. Typical embankment cross section, with rockfill downhill and reinforced earthfill uphill

bearing capacity and negligible deformation and settlements. The schist formation has a petrologic sequence of thin-bedded cherts, prevail with intercalations of clay schists, volcanic rock masses, while siltstones, sandstones and limestones were also locally included. The upper weathered part of those formations and the alluvial zone was removed before the embankment construction.

The embankment slope angle, for the rockfill was 2:3 (34°, vertical: horizontal) and for the reinforced earthfill was 2:3 to 1:1 (34° to 45°, vertical: horizontal). Berms 5-10-20 m high, with horizontal benches ≥ 4 m, were shaped at several levels, in order to obtain satisfactory factors of safety in the slope stability design.

In order to increase the shear strength and the slope stability conditions, both in static and in seismic loads ($\alpha = 0,30$ g), only the earthfill was reinforced by Geogrids of nominal tensile strength 55-80-110 kN/m, in lift layers every 600 mm, wrapped around 3 m at the sections that the earthfill had a slope inclination of 1:1 (45°). In order to separate the rockfill from the earthfill, a 300 gr/m² non-woven, needled punched, geotextile was placed. Rock boulders were used at the fill toe, for slope and for erosion protection. For the earthfill facing protection against erosion, a FORTRAC 3D 90 geogrid was placed.

The typical embankment cross section is given at Fig. 3. The design parameters concerning the basic physical and mechanical properties of the fill and the ground materials, are summarized at Table 4.

6 Conclusions

Since the wheel invention into Neolithic period, rock materials ("λίθος", "lithos") were used for road construction. Roadways of the classical era consisted of angular stones, establishing pavements of significant strength.

Scientific knowledge combined with applied technology, allows today the use of the available materials, either the earthfill and the rockfill materials, in the road embankment construction. Although earthfill is widely known and specified in detail as an embankment material, rockfill enables road designers and contractors to construct stronger and safer geostructures. Moreover, recent experience from motorway construction projects proves that rockfill embankments may attain significant heights and exhibit insignificant deformation and subsidence, under traffic loads and design loads.

Specific requirements concerning the origin and strength of the available materials, the physical and strength characteristics, as well as construction guidelines, are essential, prerequisites for a successful application of rockfill to building durable motorway embankments.

References

1. EN ISO 14688-1 to 2. Geotechnical investigation and testing - Identification and classification of soil - Part 1: Identification and description - Part 2: Principles for a classification
2. EN ISO 14689. Geotechnical investigation and testing - Identification and classification of rock - Part 1: Identification and description
3. prEN 16907-1 to 6. Earthworks
4. US Department of Agriculture, Part 645 Construction Inspection National Engineering Handbook, Chapter 8: Earthfill and Rockfill
5. ISSMGE. Geotechnics in pavement and railway design and construction. Proceedings of Int. Seminar in Athens (2004)
6. Burland, J., Chapman, T., Skinner, H., Brown, M.: ICE manual of geotechnical engineering (2012)
7. Look, B.: Handbook of Geotechnical Investigation and Design Tables. Taylor & Francis, Balkema (2007)
8. Leps, T.: Review of shearing strength of Rockfill. Journal of SMFD, Proceedings of ASCE (1970)
9. Smoltczyk, U.: Geotechnical Engineering Handbook. Volume 2: Procedures, Ernst & Sohn (2003)
10. Mouratidis, A.: Construction and performance of rock embankments in highway engineering. In: Proceedings of 4th International Conference on Bituminous Materials and Mixes, Thessaloniki (2007)

Slope Movements in the Southwest of Khoy County, Iran (Characterizations, Causes and Solutions)

Akbar Khodavirdizadeh[1](✉) and Ebrahim Asghari-Kaljahi[2]

[1] Engineering Geology, Khoy, Iran
khodaveirdizadeh@yahoo.com
[2] Department of Earth Sciences, University of Tabriz, Tabriz, Iran
e-asghari@tabrizu.ac.ir

Abstract. The southwest area of Khoy County, in the Northwest of Iran, due to geological conditions and its location in Sanandaj-Sirjan geological zone, is one of the hazardous regions regarding to landslides. Geology formations of this area are consisting of sedimentary, metamorphic and volcanic rocks. In this paper, first, the geological characterizations of the study area have been presented, and then the instable slopes are introduced. The reasons of slopes instability have been discussed. The main factors of slope instabilities in this area are ground-water, rock weathering, dynamic shocks and road/railway construction are known. In this paper, the conditions and factors (characterizations) of landslide activity have been analyzed. In south west of Khoy, some landslides in road slopes occur after heavy rainfalls. There are some historical landslides along roads slopes and rural residential areas. Main historical landslide is Gougerd in the south of the Avrin Mountain. The mechanism of most landslides in soil slopes are rotational, the depth of slide surface is ranging from 3 to 45 m. The height road slope slides reaches to 50 m and in mountain areas reach 1100 m. The results showed that the main factors of landslides are precipitation and groundwater table rising, erosion of slope foot by the rivers and dynamic loads such as of railway and road traffic.

Keywords: Road slopes · Landslide · Groundwater · Gougerd · Khoy

1 Introduction

One of the important geological hazards in the Khoy County, in the northwest of Iran, is landslide. Due to the effect of tectonic forces, the sediments of the area are crushed and have relatively large junctions, which sometimes lead to re-activation of landslides by manipulation and loading. In general, the intrinsic properties of rocks and the levels of fractures and the high rainfall are the main reasons for the slope instability of this region. In this study, in addition to reviews the characteristics of movements, the cause and prevention of this phenomenon area is studied. Figure 1 shows the position of the study area. Landslide is one of the oldest and most dangerous phenomenons in this

© Springer Nature Switzerland AG 2020
H. Shehata et al. (Eds.): GeoMEast 2019, SUCI, pp. 94–103, 2020.
https://doi.org/10.1007/978-3-030-34184-8_6

area. To continue research on the landslide, not a long time that this belief has been created between in the residents people of this area that the landslide occurrence after each heavy rainfalls (Fig. 2).

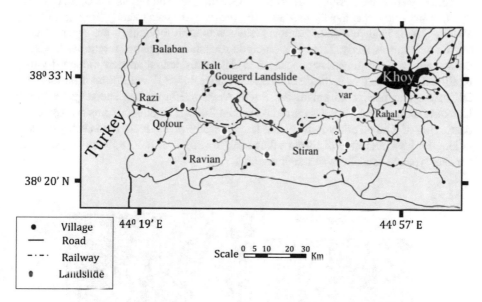

Fig. 1. Location of study area in the north west of Iran and existence landslides

A. In Near Hot Water Springs (Near Shah Abbasi Caravansary), In June 2019, View to Eastward

B. The Khoy-Türkiye Road & Gotour River, In 2013 View to West

Fig. 2. Some views of road slopes in the southwest of Khoy

2 Methods and Approaches

The results of field studies based on the engineering geology studies on the hazards of landslides were done during 2007 to 2019 (Khodavirdizadeh et al. 2010, 2011, 2012, 2014). This study was mostly based on the process of landslides by Terzaghi (1950). Varnes (1978) landslide classification system was used to identify the types of slope movements in this study. Landslides include rock falls, toppling, rotational landslide, creeps activities and flows. Amongst landslides in residential areas is Gougerd village rotational landslide that is located in east longitude 44°, 34' and northern latitude 38°, 29' at the distance of 45 km in the South west of Khoy city. The major effect of landslide on the road slopes occurred 3 August 2009. The effect of this landslide was destroying of Khoy-Turkey railway at a length of 300 m and main road (Khoy-Turkey) at length of 1 km around of Stiran village. Figures 3, 4, 5, 6, 7, 8, 9 and 10 show the various types of landslides in the study area.

Fig. 3. Two pictures that show Khoy-Turkey road demolition by a major landslide

Fig. 4. Some pictures that show rock falls in Khoy-Turkey road

G₁.

Rock fall on the Khoy-Türkiye Road
In 2013, View to West

G₂ G₂.

Rock fall on the Khoy-Türkiye Road
In 2016, View to West

Fig. 5. Two pictures of the rock fall on the Khoy-Turkey road

H. Toppling Hazard Probability in Khoy –
Türkiye Road
June 25, 2010, View to West

Fig. 6. Some pictures of the toppling hazard on the Khoy-Turkey road

June 20, 2019, View to North East

J. Toppling Hazard Probability in Khoy-
Türkiye Road

Fig. 7. Some pictures of toppling hazard on the Khoy-Turkey road

K. Observation of Creep Event on Khoy Railway Slope, In 2009

L. Observation of Creep Event in Khoy-Türkiye Road, In June 20, 2019, View to East

Fig. 8. Some pictures that show of the creep activities in Khoy-Turkey road

Effects of Landslide Event in Gougerd Village – Khoy
M. In April 2, 2013 - View to North East
N. In 2008 - View to West

Fig. 9. Some pictures that show of the effect of landslide activities in rural residential areas

A View of Gougerd Landslide

O. Photograph of Gougerd Village, In June 25, 2010 – View to West
P. Satellite Image, Google Earth, in 2014 - View to North

Fig. 10. Some views of the effect of major landslide in Gougerd area

3 Geological Studies

Based on structural-sedimentary classification of Iran, the study area is located in zone of Alborz-Azerbaijan. In this area sedimentary, metamorphic, volcanic rocks of last Precambrian to recent deposits have outcrop. The rock units are include pieces of calcareous, conglomerate, sandstone, ultramafic rocks and tuff, shale, schist and clay, (Geological survey of Iran 1978). The tectonic of the study area is on the Khoy Ophiolite by tectonic forces converging in the phase of the Laramide. By tectonic forces time equivalent the Laramide and convergence the Iran-Arabia, block and thrust faults proceeds, in the upper cretaceous and Paleocene have been raised. Tectonic forces with by creating folds and faults direction from northwest to southeast

Q. Wash and Erosion the Foot of the Road Slopes By the Gotour River, in 2012

Q₁. View to North East
Q₂. View to West

Fig. 11. Some pictures that show of the effect of Qotour River erosion on the foot of the road slopes

R. A View of the High Groundwater Surface In the Khoy - Türkiye Road, View to West

Fig. 12. Some pictures that show the effect of the high groundwater table as the cause of landslide

characterized in area. Topography of area full rugged whit the slope was steep and contains multiple channels is along the north-south. Due to the weakness and high erosion Ophiolite series in area, in the surface of slopes a relatively thick layer of weathered material is recognizable (Geological survey of Iran 1978) (Figs. 11 and 12).

Hydrology of study area shows that due to heavy rainfall, materials slopes water-saturated and therefore water penetration destructive factors and increased in slope movements. According to the rains on slopes crown from inside the slopes soil materials to the seasonal rivers slope in downstream (Qotour River and Avrin Rivers) is drainage. In the slope slide also large amounts of precipitation and intruded into slopes. And slope soil mass is saturated. This causes the accumulation and storage of water in the lower parts and the reduction of soil strength and increases the instability of slopes. Qotour River located in Aras River catchment. And watershed area it has been created folds of the Northwest Zagros Mountains of and this river to plain from of high mountains length of 70 km is formed (Armed Forces Geographical Organization of Iran 1998; Geological survey of Iran 1978).

4   Landslide Characterizations and Effects

There are many small and large landslides in the study area. The important landslides and slope instability are as follow:

4.1   Gougerd Landslide

Gougerd village landslide located in latitude $38°$, $29'$ and longitude $44°$, $34'$ in 45 km southwest of Khoy City. This landslide is located on hill slide at west with angle dip to 18 at 500 m height from main road and limited by the small Rivers in the west and south (Figs. 9 and 10). Landslide with area 150 hectares and the high different of top and foot is about 230 m. The main mechanism of Gougerd village landslide is rotational and mud-flow. Gougerd village houses are developed on the central part of this landslide (Fig. 10). Landslide activity caused some fissures on the walls and foundations of buildings (Fig. 9). A landslide dimension of agricultural land is about 100 m and in length 50 m. Average height difference of top to toe of slide is about 30 m. Distance from the landslide crown is South west of the village (houses are in South part of the village) is about 100 m. If landslide phenomena continue, village houses will damage. Frequency of tensile cracks in the slip and slide on top of the crown is visible. The material of slope in surface layer is silty sand (SM) to a depth of 3 meters and then is clayey gravel (GC). Village houses were built on the central part of landslide. Slope stability problems in the Gougerd village, according to the residents of the village, about 20 years ago, has been created and after each heavy rainfall has exacerbated the problem (Figs. 9 and 10).

Based on geological survey, slide mass is consisted of alluvial terraces, alluvial fan and alluvial river bed. Due to developing of Gougerd village on the Ophiolitic rocks, weathering and producing of sediments are common. Also the deposits of old big landslide are loose and instable. Lithology of the slide mass in the west Gougerd village affected from this two phenomenon. Considering rocky pieces inside the slide mass is angular and few eroded, this possibly is main origin soil slide mass refer old landslide occurrence and or that alluvial fan, whereas short distance transportation. Also in nearness the river, rock fragments relatively erosion therefore very highly possibly river origin. Generally the slid mass came from three origins; old major landslide, alluvial fan-old alluvial terraces and river sediments (Khodavirdizadeh et al. 2014). Considering the Gougerd village is situated on sediments of on old landslide in the slope hillside, the main cause of the landslide on the this slope is seepage much amounts of water from upstream village masses into the soil layers and drainage which is sliding toward the slope toe. The causes of rising groundwater level and pore pressure increase in mass and this is very impressive and significant factor in exacerbating or even starting to slide. Ascension groundwater table and saturation of slopes and increase pore pressure and load are main factors of Gougerd landslide (Khodavirdizadeh et al. 2014, 2012, 2011, 2010). For the Gougerd village landslide, presentation of water penetration from upstream (crown of landslide), groundwater drainage, removing soil from top parts are recommended for stabilizing.

4.2   Mud Flow of Gougerd Area

This landslide is old mud flow with area about 1000 hectares and the different of top and foot is about 1100 m, with variable slope between 25 and 45°. Direction of Gougerd landslide is a variable and is limited by the channels and the eradication the village is the eradication of upstream and demolition of rural houses on old landslide construction has been developed. Relatively deep channels on both sides of landslide developed. And the surface water drainage takes place through them. The main slide area is located in the village location that due to low permeability sediments and soil type and increased pore pressure in saturated soil conditions to induce masses slide provides the location and surface area of above factors makes up a lot of time for water to penetrate to the masses, most of the surface water.

According to Cruden and Varnes (1996), description of flow, in Gougerd area a flow in type of mud flow movement detected on a very large scale. "Mudflow is a very rapid to extremely rapid flow of saturated plastic debris in a channel, involving significantly greater water content relative to the source material (Plasticity index > 5%)" (Hungr et al. 2001).

4.3   Landslides in Road Slopes

Based on field studies, the main mechanism of these landslides is rotational. The height difference of crown to toe up 50 m for these landslides was determined. Slide direction of these slides along the road slopes are south and toward to Qotour River. The effects of these landslides are, in addition to the destruction of the main road, sedimentation inside the Qotour River. Most landslides are associated with creeps, including road slopes and railway slopes. And there is a meaningful relationship between them. In fact, it can be said that all the rock movements in the road slopes are due to the continuous activity of this phenomenon. For this phenomenon the height difference of crown to toe up in slopes is 50 m. creeps are seen in many parts of the study area, especially along the main road of Khoy-Turkey. Although the groundwater table changes due to cause of road slope landslides as the only cause of landslides is not considered but for the occurrence of landslides on the slopes of the road are considered a strong main factor. An acceptable relationship also the occurrence of landslides on the road slopes with railway track across slopes wash the toe and continuous erosion of mentioned the slopes by continual Qotour River. Moreover vibrations of the train on the railway track, earthquake and the type of slope materials also have an active role in the occurrence and progression of landslides.

For protection and stabilization of road slopes are recommended using wire mesh chain for rock falls, using gabion wall at the toe of the slopes along Qotour River and retaining wall for prevention of erosion in toe slopes.

4.4 Rock Falls

Rock falls caused damages and destruction into the slopes main road Khoy-Turkey (Figs. 3 and 4). These movements are caused by erosion of rocky crags and any type loading too much at the top of the slope or by cutting the base of the slopes (Varnes 1996). Rock falls after downfall can cause large-scale destruction of the main road and they are blocked. Rock falls also after each downfall and destruction occur in near slopes. This phenomenon also can be source of vibrations (transportation of cars on the road and railway) or local earthquakes. Rock falls often associated with landslides on the slopes of the road and downfall into the Qotour River, resulting in devastating damage and erosion of slopes.

4.5 Topplings

The risk of this phenomenon is more evident on the main road. The danger of active this phenomenon on the main road is more evident. And these phenomena are only on the right side of the road towards Turkey. Vertical cracks and fractures in the masses of road slopes that can be caused by numerous geological factors is the main cause of the toppling dangerous possibility of road slopes.

5   Conclusions

The results of studies and analysis on the slope movements of southwest of Khoy County show that the main reasons of these movements are as follow:

- Topography,
- Precipitation and rising of the groundwater table,
- Erosion of slope foot by the rivers,
- Dynamic forces such as vibration of railway/road traffic.

There are various types of slope instability for rock and soil slopes of the study area. Rock fall, toppling, rock slide and rotational sliding of soil slopes and mud-flow are the common types of slope movements. Slope movements affect the international road and railway of Iran-Turkey, rural residential and agricultural lands.

One of the large slope movements is Gougerd landslide. Having 150 hectares area and 230 m, the main mechanism of Gougerd landslide is rotational and mud-flow. Gougerd village houses are developed on the central part of this landslide. Gougerd landslide activity caused some fissures on the walls and foundations of buildings of Gougerd village.

For protection and stabilization of road and railway slopes are recommended groundwater drainage, using wire mesh chain for rock falls, gabion wall at the toe of the slopes along Qotour River and retaining wall for prevention of erosion in toe slopes.

References

Armed Forces Geographical Organization of Iran: Khoy Topographic Map 1:250,000 Temper Series k551, 2nd edn. Leaves NJ38-6 (1998)

Cruden, D.M., Varnes, D.J.: Landslide types and Processes, Transportation Research Board, U.S. National Academy of Sciences, Special Report, vol. 247, pp. 36–75 (1996)

Geological survey of Iran: Geology Map of Khoy, Scale, 1:250,000 (1978)

Hungr, O., Evans, S.G., Bovis, M., Hutchinson, J.N.: Review of the classification of landslides of the flow type. Environ. Eng. Geosci. 7, 221–238 (2001)

Khodavirdizadeh, A., Asghari-Kaljahi, E., Abolhasanzadeh, S.: Sensitivity of soil cohesion on the stability of gougerd landslide, Northwest of Iran. In: Engineering Geology for Society and Territory, vol. 2, pp. 1281–1284. Springer (2014)

Khodavirdizadeh, A., Asghari-Kaljahi, E., Abolhasanzadeh, S.: The effect of soil parameters on the stability of Gougerd village landslide. In: 2nd National Conference on Structure, Earthquake and Geotechnics, Babolsar, Iran (2012). (in Persian)

Khodavirdizadeh, A., Razizadeh, F., Asghari-Kaljahi, E., Poormohsen, M.: Pseudo-Static stability analysis of Gougerd landslide under different groundwater conditions. In: First International Conference on Urban Construction in the Vicinity of Faults (ICCVAF 2011), Tabriz, Iran, 3–5 September 2011. (in Persian)

Khodavirdizadeh, A., Razizadeh, F., Poormohsen, M.: Cause of slide investigation and static stability analysis of Gougerd landslide in the Khoy area. In: The 14th Symposium of Geological Society of Iran and 28th Symposium on Geosciences, 16–19 September 2010, Urmia University (2010). (in Persian)

Terzaghi, K.: Mechanism of landslides. In: Engineering Geology (Berkel) Volume. Ed. da The Geological Society of America, New York (1950)

Varnes, D.J.: Slope movement types and processes. In: Schuster, R.L., Krizek, R.J. (eds.) Landslides Analysis and Control, Transportation Research Board, Rep. No. 176, National Academy of Sciences, pp. 11–33 (1978)

Human-Induced Landslide Risk Analysis at Durhasanlı Reservoir Site

Murat Mollamahmutoğlu[1](✉) and Yılmaz Acar[2]

[1] Engineering Faculty, Civil Engineering Department,
Gazi University, Ankara, Turkey
mmolla@gazi.edu.tr
[2] Geolimit, Ankara, Turkey

Abstract. In this research, it was aimed to investigate the potential landslide induced by spillway construction activities at Durhasanlı reservoir site located in Manisa province of Turkey and provide a feasible stabilization measure. Since the site reconnaissance revealed that some deformations on spillway concrete members and the development of tension cracks at the head of hillside on the right bank of spillway, five inclinometers were immediately placed at critical points and the possible slope movement was observed for 40 days. It was found out that the mass movement were going on at a rate of about 0.1 mm per day. With the help of inclinometers' data, the depth of sliding mass was also determined. Accordingly, based on geotechnical properties obtained from surface and subsurface investigations, a number of slope stability analyses were carried out taking into account such factors as the tension cracks, earthquake loading, reduction of slope weight and pile reinforcement. As a result, the secured solution for the stability of mobilized slope was suggested by terracing the hillside, pile reinforcement, and effective drainage system.

Keywords: Landslide · Tension crack · Pile reinforcement · Slope stability analysis

1 Introduction

Landslides pose a big threat to humankind, natural environment, constructed facilities, and infrastructures. They are natural phenomena controlled by gravity but they are sometimes triggered by such external factors as rainfalls, earthquakes, volcanic eruptions, floods, and human activities. The population increase and urbanization intensify the alteration of the landscape, which can be a contributing factor in causing landslides. Many human-triggered landslides can be avoided or mitigated. They are commonly due to the construction of civil or industrial structures, bridges, roads, tunnels, pipelines, water artificial reservoirs, deforestation of land, incorrect agricultural melioration works, worsening of flow-off surface and underground waters and failures of the underground pipelines. Several catastrophic landslides over the last 100 years have been caused by human carelessness (Muller 1968; Guadagno et al. 1999; Sammarco 2004) when modifying the landscape. Based on the investigation of landslides in Slovakia it was stated that 90% of all landslides, which occurred in populated areas in

© Springer Nature Switzerland AG 2020
H. Shehata et al. (Eds.): GeoMEast 2019, SUCI, pp. 104–115, 2020.
https://doi.org/10.1007/978-3-030-34184-8_7

the last 30 years and which caused the greatest damage, were brought about fully, or partly by man's intervention (Malgot 1980). Nilsen and Turner (1975) stated that 80% of the landslides, which had taken place since 1971 in California, were brought about by artificial interventions. Further interesting examples of human impact on landslides were given by Sassa (1999).

The most common human-triggered landslides are induced by cut and fill for road constructions. This is especially true in Southeast Asia where the development of roads in mountainous areas using fill slope methods has led to a significant number of debris-flows (Collins 2008). The construction of highways resulted in an increase in landslide in the Garhwal Himalaya (Barnard et al. 2001). The formation of new slope such as in mining operations is well-known trigger of landslides. One of the famous ones is the Aberfan disaster in Wales in 1966, which killed 144 people (Siddle et al. 1996).

Landslides are commonly triggered by the modification of slope stability conditions, increasing stress, or reductions in strength (Rahardjo et al. 1988; Ng et al. 2003, Ng et al. 2008). Loading a slope can trigger a slide, and the most famous example is the Rissa landslide in Norway (Gregersen 1981).

In the case of surface water flow modifications, soil sealing induced by surface changes such as road, parking, etc. can be triggering and degrading factor. In Menton area (France), a heavy rainfall in November 2000 induced 400 landslides. The main causes were the urbanization that led to runoff flow path getting out of control, created by new roads, terraces, etc. (Safeland 2011).

In this study, the possible landslide caused by spillway construction activities at Durhasanlı reservoir site was closely monitored by means of inclinometers for a while and meanwhile, the required stability analyses were carried out to come up with a best possible corrective measure before the failure took place.

2 The Study Area

Durhasanlı reservoir site is located in Manisa, Turkey's Aegean Region. The geology of the study area and its immediate vicinity is formed by Paleozoic aged Eşme formation composed of the metamorphites of Menderes Massif. Tertiary, quaternary and senozoic aged transitive sedimentary and volcanic rocks takes place uncomfortably over the metamorphic. Terrestrial deposits which belong to mid-Miocene were present from place to place. Kula volcanites exist in the region at most. The quaternary aged hill rubbles and alluvial units were situated at the top. The Eşme formation is gray-fawn colored and includes schist fines such as augen-gneisses, biotite-gneisses with quartz veins and mica schist, quartz, muscovite schist, chlorite schist, sericite schist and biotite schist from bottom to top. The units which belong to Eşme formation are highly affected by the local tectonism and there exist micro dimensional folding. Moreover, anticlinal structure is found in the reservoir area.

Although outcropping schists are folded due to tectonism no faulting is encountered during site investigation. Nevertheless, the closest active faults belonging to Alaşehir graben fault systems are found in the south of the study area and therefore the study area is regarded within the highest seismic zone according to Turkey's earthquake zonation map.

3 Methodology

The method included the surface and subsurface investigations, in-situ and laboratory tests, slope monitoring, slope stability modeling, and analyses.

3.1 Surface and Subsurface Investigations

During the beveling excavations for Durhasanlı reservoir spillway, some earth flows took place just behind the right bank of spillway. Meanwhile, some deformations were observed on the reinforced concrete members of spillway. This occurrences provoked anxiety that there might have been the process of imminent landslide. Hence, the scrutiny of the area under consideration revealed that tension cracks materialized at the head of hillside (Figs. 1 and 2).

Fig. 1. The location of tension cracks

Fig. 2. Appearances of tension cracks at the location

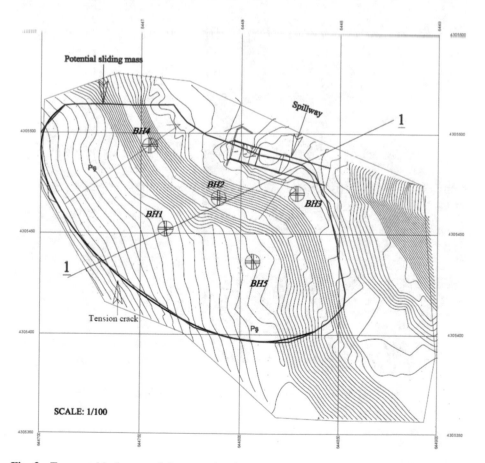

Fig. 3. Topographical map of the area showing the potential sliding mass boundary and the boreholes' locations.

3.2 Slope Monitoring

Within mobilized area, 5 boreholes, each of which had about 40 m boring depth, for inclinometer measurements were arranged without delay and the boreholes were located downslope at critical points as shown in the topographical map of the area (Fig. 3).

Having completed the drilling works, the boreholes were installed with inclinometers' pipes of 76 mm in diameter and the readings were taken right after for a period of 40 days. The inclinometer data from each borehole indicated that there was a creep of the mass under consideration at a rate of about 0.1 mm/day. The movement depths of sliding mass were also determined for boreholes of 1, 2, 3, and 5. They were 21 m, 14 m, 10 m and 16 m respectively, which enabled the potential slip surface to be set for the slope stability analysis. The typical borehole readings, the depth of movement and the corresponding soil materials (Çakır et al. 2018) were illustrated in Figs. 4, 5 and 6.

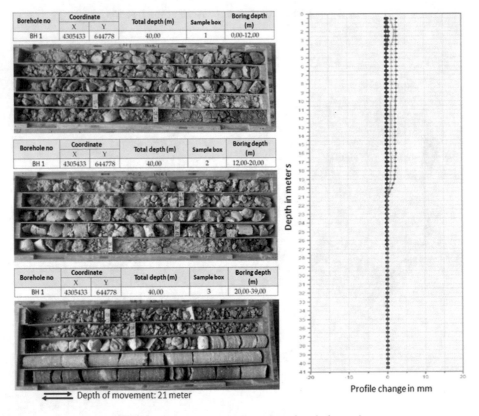

Fig. 4. Inclinometer readings from borehole no 1

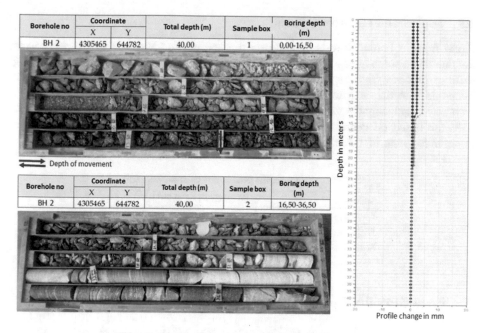

Fig. 5. Inclinometer readings from borehole no 2

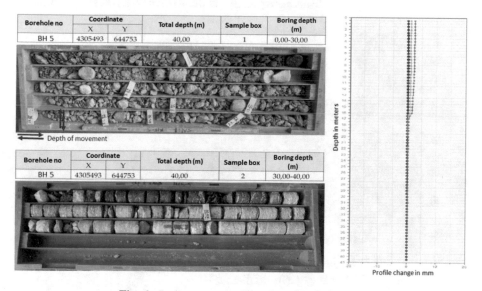

Fig. 6. Inclinometer readings from borehole no 5

3.3   In-Situ and Laboratory Tests

Apart from inclinometer tests, identification, classification and unit weight tests were conducted on some representative disturbed samples of landslide material taken during drilling process. Since it was not possible to obtain undisturbed specimens from landslide material and weathered schist, the shear strength parameters of highly weathered and slightly weathered schists were determined by means of RockLab software based on generalized Hoek-Brown failure criteria and the results for each layer were given in Table 1 below.

Table 1. Engineering properties of rock and soil

Soil/Rock	γ (kN/m^3)	C (kPa)	ϕ (Degree)
Landslide material (GW-GC)	18	23[*]	9[*]
Highly weathered schist	20	30	15
Schist	24	100	35

[*] obtained from back analysis

4   Slope Stability Modelling and Analysis

The most critical cross-section (section 1-1) of the hillside under consideration was obtained from Fig. 3. In addition, based on data procured during surface and subsurface investigations, the slope for the necessary analyses was then modeled as shown in Fig. 7.

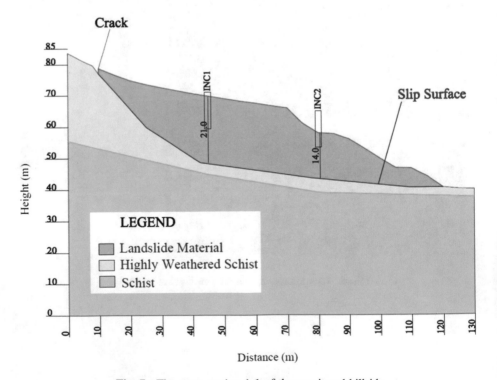

Fig. 7. The cross-section 1-1 of the monitored hillside

The top landslide material (silt, sand and gravel sizes resulted from the disintegration of schist) was successively followed by highly and slightly weathered schist layers and no groundwater was encountered up to the drilling depth.

For the slope profile constructed in Fig. 7, the back analysis was carried out by Slide 7 to determine the shear strength parameters of landslide material presented in Table 1 as well as the tension crack depth which was about 3 m (Fig. 8).

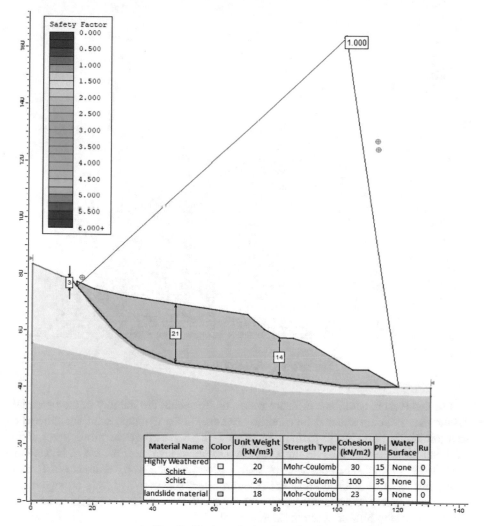

Material Name	Color	Unit Weight (kN/m3)	Strength Type	Cohesion (kN/m2)	Phi	Water Surface	Ru
Highly Weathered Schist	☐	20	Mohr-Coulomb	30	15	None	0
Schist	☐	24	Mohr-Coulomb	100	35	None	0
landslide material	☐	18	Mohr-Coulomb	23	9	None	0

Fig. 8. Back analysis of section 1-1

In the stabilization of the potential landslide, the first and economical attempt was to reduce the weight of sliding mass. In this regard, the removal of soil was made by terracing the hillside as much as possible (about 31% of the potential sliding mass) as shown in Fig. 9. Nevertheless, the factor of safety was less than 1.50 even under static condition.

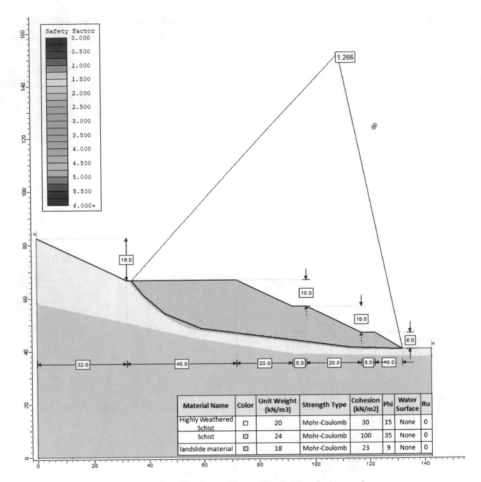

Material Name	Color	Unit Weight (kN/m3)	Strength Type	Cohesion (kN/m2)	Phi	Water Surface	Ru
Highly Weathered Schist	☐	20	Mohr-Coulomb	30	15	None	0
Schist	☐	24	Mohr-Coulomb	100	35	None	0
landslide material	☐	18	Mohr-Coulomb	23	9	None	0

Fig. 9. The proposed shape of hillside after terracing

For this reason, additional measure was taken to secure the stability of the potential sliding mass not only under static condition but also under dynamic condition since the reservoir region was in the earthquake zone I (highly earthquake prone area). The additional measure was the pile support at two different level near the toe of hillside as illustrated in Fig. 10. The piles were arranged as tangent pile with a diameter of 1 m to prevent the flow of landslide material through piles' spacing. With this arrangement, the factor of safety was found to be 1.182 higher than 1.15 for earthquake condition and therefore, the hillside was considered stable.

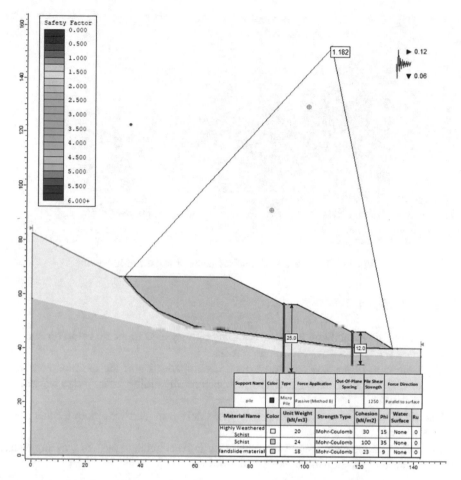

Fig. 10. Stabilization of hillside by pile support at two different level

Moreover, the horizontal deformations of stabilized slope during earthquake was also assessed by Plaxis 2D software and the maximum horizontal displacement was found to be about 12 cm localized around the upper row piles without collapse (Fig. 11).

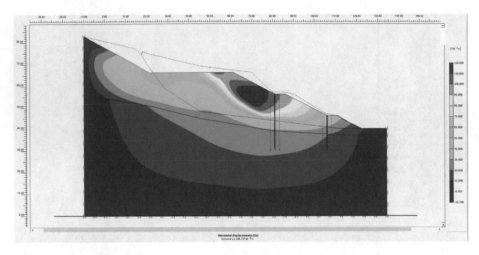

Fig. 11. The horizontal displacements of stabilized hillside

5 Conclusions

1. The potential landslide on the right bank of spillway at Durhasanlı reservoir site was triggered by human construction activities.
2. With the help of surface and subsurface investigations and the inclinometer measurements, the boundaries and the dimensions of sliding mass were accurately established.
3. The stabilization of potential landslide was safely provided with both terracing and pile support.
4. In addition to the stabilization methods, an efficient drainage system was suggested in and around the hillside under consideration.

References

Gregersen, O.: The quick clay landslide in Rissa, Norway. In: Proceedings of the X International Conference on Soil Mechanics and Foundation Engineering, Stockholm, Sweden, vol. 3, pp. 421–426 (1981)

Collins, T.K.: Debris Flows caused by failure of fill slopes: early detection, warning, and loss prevention. Landslides **5**, 107–120 (2008)

Muller, L.: New considerations on the Vajont slide. Int. J. Rock Mech. Min. Sci. **6**, 1–91 (1968)

Guadagno, F.M., Celico, P.B., Esposito, L., Perriello, Z.S., Piscopo, V., Scarascia-Mugnozza, G.: The debris flows of 5–6 May 1998 in Campania, Southern Italy. Landslide News **12**, 5–7 (1999)

Sammarco, O.: A tragic disaster caused by the failure of tailing dams leads to the formation of the Stava 1985. Found Mine Water Environ **23**, 91–95 (2004)

Nilsen, T.H., Turner, B.L.: Influence of rainfall and ancient landslide deposits on recent landslides (1950–51) in urban areas of Contra Cose County. Bull. US Geolog. Surv. **1399** (1975)

Sassa, K.: Landslides of the world. Kyoto University Press, Kyoto (1999)

Malgot, J.: Vplyv antropogénnych faktorov na stabilitu zosuvných území na Slovensku. Geologický průzkum **22**, 139–143 (1980)

Rahardjo, H., Leong, E.C., Gasmo, J.M., Tang, S.K.: Assessment of rainfall effects on the stability of residual soil slopes. In: Proceedings of the 2nd International Conference on Unsaturated Soils, Beijing, China, vol. 2, pp. 280–285 (1988)

Ng, C.W.W., Zhan, L.T., Bao, C.G., Fredlund, D.G., Gong, B.W.: Performance of an unsaturated expansive soil slope subjected to artificial rainfall infiltration. Géotechnique **53**, 143–157 (2003)

Ng, C.W.W., Springman, S.M., Alonso, E.E.: Monitoring the performance of unsaturated soil slopes. Geotech. Geol. Eng. **26**, 799–816 (2008)

Barnard, P.L., Owen, L.A., Sharma, M.C., Finkel, R.C.: Natural and human-induced landsliding in the Garhwal Himalaya of Northen India. Geomorphology **40**, 21–35 (2001)

Siddle, H.J., Wright, M.D., Hutchinson, J.N.: Rapid failures of colliery spoil heaps in the South Wales Coalfield. Q. J. Eng. **29**, 103–132 (1996)

Nadim, F., Høydal, Ø, Haugland, H., McLean, A.: Analysis of Landslides triggered by anthropogenic factors in Europe, SafeLand European Project Living with Landslide Risk in Europe: Assessment, Effects of Global Changes, and Risk Management Strategies, 81 p. (2011)

Çakır, E., Acar, Y., Özarslan, M.: Geotechnical report for the right bank landslide of Manisa-Demirci-Durhasanlı reservoir, Geolimit, Ankara, Turkey, 32 p. (2018)

Application of Bottom Ash as Filter Material in Construction of Dyke Embankment for Sustainable Infrastructure

Vinod Kumar Mauriya[✉]

PE-Civil, NTPC LTD, Gautambudh Nagar, Noida, India
vkmauriya@ntpc.co.in

Abstract. India's present energy need from coal reserves contribute about 55% of the total power generation. Indian coal has high ash contents and presently about 200 million tonnes of ash is generated annually. The generated ash has become a subject of worldwide interest in recent years because of its diverse uses in manufacture of cement, bricks and concrete block, filling of underground cavities, mine voids etc. The present ash utilisation is about 69% and the balance ash is stored safely in ash ponds. The utilization of ash is strongly related to economy of its use, which changes with time. There may not be users in sight today, but after some time there could be bulk users.

The most economic and commonly used method to dispose ash is by hydraulic transport, in the form of slurry, to the ash pond. At the ash pond, storage space is created by constructing dyke embankments all around, within which ash particles will be allowed to settle and the decanted water is allowed to escape for recirculation back to plant. Based on the type of the soil available for the embankment construction, mostly, it is a homogeneous section with internal drainage arrangement of sand chimney and sand blanket. The dyke embankments are designed as water retaining structures as per relevant Indian Standards applicable for earth dams. For internal drainage in these embankments, natural river sand or crushed stone sand is generally used. To create sustainable and environment friendly infrastructure, NTPC has recently explored the use of bottom ash as filter material with respect to pond ash as base material. Based on laboratory tests, it is found that bottom ash possess the required filter ability, internal stability, drainage capacity, self healing properties and does not segregate. Accordingly, bottom ash has been adopted as an alternate filter material in internal drainage system of dyke embankments of various plants of NTPC. The above have been successfully implemented and found to be safe under operating phase. This use of bottom ash as a filter material has also saved the time and cost over runs and most importantly is echo- friendly.

This paper highlights the use of bottom ash as filter material in water retaining structures such as dyke embankments/earth dams in a sustainable manner to entire world.

Keywords: Dyke embankment · Bottom ash · Filter material · Base material · Raising

© Springer Nature Switzerland AG 2020
H. Shehata et al. (Eds.): GeoMEast 2019, SUCI, pp. 116–124, 2020.
https://doi.org/10.1007/978-3-030-34184-8_8

1   Introduction

Ash is a non-degradable, non-perishable, inert material, which could be used even after hundreds of years. When coal can be used after millions of years of its formation, its by-product can also be used after a long time gap. The ash generated from the power plants is disposed-off in the Ash ponds. Unlike water reservoir, the dyke embankments for ash pond are generally not constructed upto ultimate height in one go and initially constructed upto a limited height with provision of subsequent raising as per requirement. Ash pond is divided into lagoons and provided with garlanding arrangements for changeover of the ash slurry feed points for even filling of the pond and for effective settlement of the ash particles. A photograph of actual ash pond of NTPC with ash discharge line is presented in Fig. 1.

Fig. 1. Photograph of actual ash pond of NTPC with ash discharge line

Ash pond is designed as multi-lagoon systems with minimum two storage lagoons and one over-flow lagoon (OFL). Having two or more storage lagoon facilitates sequential raising of lagoons by putting one lagoon for ash filling while the other lagoon is used for raising its dyke. OFL helps in controlling the effluent quality of the supernatant, which is recycled back to plant for making ash slurry. The typical layout and cross-section of storage lagoon and OFL is shown in the following Figs. 2 and 3 respectively.

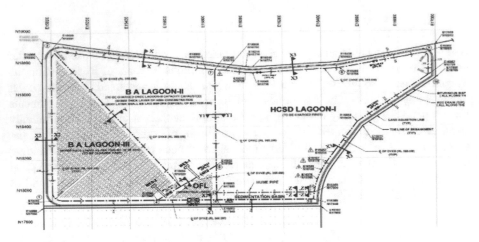

Fig. 2. Typical layout of ash pond

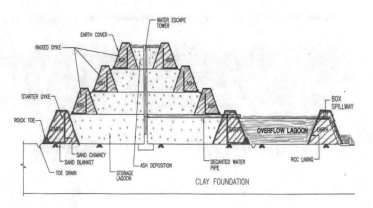

Fig. 3. Typical cross-section of storage lagoon and OFL

2 Dyke Embankment

Dyke embankment is a retaining structure to contain ash slurry (or continuous placement of unused ash to ensure uninterrupted operation of the thermal plant) and settled ash (till it is used for any beneficial purpose). The embankment of the dyke is designed with sufficient strength and safety features to avoid breaching at any point of time. The upstream slopes of the embankment is protected from wave action using brick lining and downstream slope is protected from rain cuts using stone pitching/turfing. Suitable capacity Spillways are provided to release the excess rain-water and maintain adequate free board to avoid any chance of overtopping of dyke.

2.1 Internal Drainage System

The basic requirements for design of dyke embankments are to ensure; safety against stability, safety against internal erosion and safety against overtopping. To ensure safety against internal erosion; internal drainage system is provided. A typical section of upstream raising with internal drainage system adopted by NTPC is indicated in following Fig. 4.

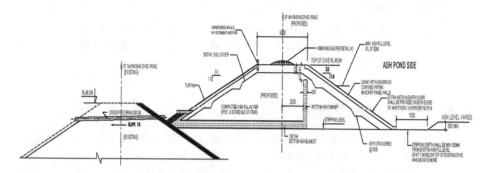

Fig. 4. Typical section of upstream raising

As the dyke embankment constructed with earth/ash is a porous structure, it allows a gradual movement of water through its pores as can be seen in the following Fig. 5(a). In order to keep the downstream slope dry and stable, internal drainage arrangement in the form of chimney and blanket are provided to intercept the seepage, if any and channelize the same through the rock toe and toe drain as presented in following Fig. 5(b).

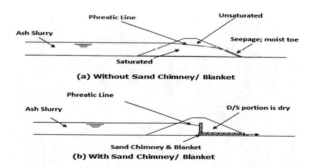

Fig. 5. Dyke embankment with and without internal drainage arrangement

2.2 Filter Material

In the internal drainage system of dyke embankments, natural river sand or crushed stone sand (M-sand) is generally used. However, to create sustainable and environment friendly infrastructure, NTPC has recently explored the use of Bottom ash in the internal drainage system of water retaining structures like dyke embankment. NTPC have got the bottom ash samples tested with respect to base material as pond ash for its various plants to explore suitability as an alternate of sand filter in dyke construction works. Out of the total ash generated, generally the percentage of bottom ash is about 20% and that of fly ash is about 80%. Pond ash is a combination of both bottom ash and fly ash and therefore contains much higher percentage of fines than bottom ash alone.

2.3 Filter Criteria

The evaluation of suitability of filter material for filter ability, internal stability, drainage capacity and self healing with respect to base material is checked in line with filter criteria specified in IS: 9429.

3 Pond Ash and Bottom Ash

Fly ash is extracted from Electro Static Precipitator (ESP), conveyed to fly ash silos, mixed at slurry tanks and pumped to ash disposal areas using centrifugal pumps. Whereas, after coal combustion, the coarser fraction of ash comes at bottom of Boiler furnace and crushed by clinker grinders to become bottom ash. These are generally in the range of sand size particles, which is very hot and therefore collected in water impounded hoppers and thereafter taken to slurry tanks and pumped to ash disposal areas using centrifugal pumps. Generally, the percentage of fines (particle size smaller than 75 μ) in pond ash vary from 40% to 85% since pond ash is expected to contain much more fines than the bottom ash. The bottom ash is coarser than the fly ash and the percentage of fines (particle size smaller than 75 μ) in bottom ash vary from 0% to 5%. The schematic view of an Indian thermal power plant indicating the process of generation of fly ash and bottom ash is shown in following Fig. 6.

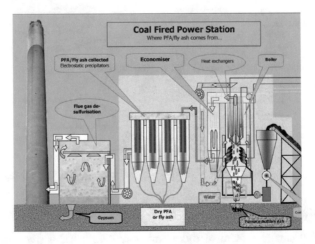

Fig. 6. Schematic view of a thermal plant indicating the process of generation of fly ash and bottom ash

3.1 Generation and Requirement of Bottom Ash

On an average, the production of bottom ash from one typical 500 MW unit is in the range of about 15,000–18,000 cum/month, depending upon the coal/combustion/heat treatment etc. The requirement of bottom ash as a filter material in raising works is not continuous and the quantity required is in the range of 50,000–60,000 cum/year for each dyke raising, which will be in the tune of 4000–5000 cum/month. The sample bottom ash is shown in Fig. 7, which is obtained through clinker grinder, the sample of which is also shown in Fig. 8.

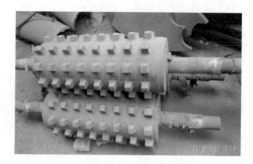

Fig. 7. Bottom ash sample   **Fig. 8.** Typ. Clinker grinder

4 Engineering Properties of Pond Ash and Bottom Ash

As per the IS Classification System, the Pond ash may be classified as equivalent to a low plastic silt or ML type soil. The lab permeability of pond ash is observed to be in the range of 10^{-5} cm/s to 10^{-6} cm/s where as that of Bottom ash is observed to be in the range of 10^{-3} cm/s to 10^{-4} cm/s. The engineering properties of pond ash (base material) and bottom ash (filter material) of some of NTPC plants is presented in following Table 1.

Table 1. Engineering properties of pond ash (base material) and bottom ash (filter material)

NTPC Plants	Pond ash					Bottom ash				
	MDD (KN/m^3)	c (KPa)	c' (KPa)	Ø	Ø'	MDD (KN/m^3)	c (KPa	c' (KPa)	Ø	Ø'
Korba	10.02	43.50	0.00	33°	33°	–	–	–	–	–
	8.4 (In-situ)	17.00	0.00	29°	30°	–	–	–	–	–
SSTPS	13.30	0.00	0.00	31°	28°	13.59	0.00	0.00	32°	31°
TTPS	11.12	20.00	–	28°	–	–	–	–	–	–

5 Evaluation of Suitability of Bottom Ash as Filter Material

5.1 Methodology for Testing

The following laboratory tests on bottom ash (filter material) as well as pond ash (base material) were carried out each on minimum five (05) samples:

   (i)   Grain size distribution
  (ii)   Density (Bulk & Dry)
 (iii)   Permeability test (Laboratory)
 (iv)   Specific Gravity
  (v)   Atterberg limits

All the above-mentioned tests were carried out in accordance with the recommendations of the relevant Indian Standards and other standard procedures. The grain size distribution is plotted on semi-log graph. The values of D_{10}, D_{15}, D_{85}, D_{90} & % passing 75 μ of the bottom ash and pond ash are worked out from grain size distribution curve for further analysis to establish the suitability of bottom ash as filter media w.r.t. pond ash as base material of dyke embankment.

5.2 Analysis of Test Results

Based on the findings of the laboratory investigations carried out on the bottom ash samples and pond ash samples, the following conclusions have been arrived at:

The grain size analysis of all tested ash samples indicate that the Bottom ash samples possess predominantly sand size particles. The grain size analysis of the Pond ash samples from the different plants indicate that the tested samples possess predominantly silt size particles. The plasticity index values of the tested ash samples indicate that all the ash samples (bottom ash and pond ash) possess non-plasticity characteristics. The laboratory permeability tests conducted on the Bottom ash samples indicate that tested Bottom ash materials possess the pervious drainage characteristics whereas as the Pond ash materials possess the semi-pervious drainage characteristics. It is observed that the material passing 75 μ sieve is less than 5% in the Bottom ash samples which meets the requirement of self-healing property necessary for filter material in the internal drainage system of dyke embankments. The Pond ash falls under category-2 of base materials with percentage finer than 75 μ between 40%–85% (as per IS: 9429). Out of the 10 plants explored, Bottom ash of the 9 plants satisfies the filter criterias. Based on the same, it is seen that Bottom ash can be used as filter material in the dyke embankments. However, the Bottom ash and Pond ash need to be regularly checked to ensure compliance of filter criterias. A typical filter material (Bottom ash) property and base soil property (Pond ash) of one of the NTPC plant is presented in following Table 2.

Table 2. Typ. typical filter material (Bottom ash) property and base soil property (Pond ash) of one of the NTPC plant

| S. No. | Location | Sampling | Material type | Specific gravity(G) | Particle size analysis | | | | D$_{15}$ (mm) | Percentage passing 75 μ |
					Clay % (< 0.002 mm)	Silt % (0.002– 0.075 mm)	Sand % (0.075– 4.75 mm)	Gravel % (>4.75 mm)		
1	POND ASH-1 (CENTRAL ASH DYKE LAGOON-1)	*DS	POND ASH	2.21	19	80	1	0	0.0015	NA
2	POND ASH-2 (CENTRAL ASH DYKE LAGOON-1)	DS	POND ASH	2.20	15	84	1	0	0.002	NA
3	POND ASH-3 (CENTRAL ASH DYKE LAGOON-1)	DS	POND ASH	2.20	14	85	1	0	0.0021	NA
4	POND ASH-4 (CENTRAL ASH DYKE LAGOON-1)	DS	POND ASH	2.21	18	81	1	0	0.0015	NA
5	POND ASH-1 (DISCHARGE LINE OF UNIT-4)	DS	POND ASH	2.24	1	2	70	27	0.18	3
6	BOTTOM ASH-2 (DISCHARGE LINE OF UNIT-4)	DS	BOTTOM ASH	2.23	1	2	77	20	0.20	3
7	BOTTOM ASH-3 (DISCHARGE LINE OF UNIT-4)	DS	BOTTOM ASH	2.24	1	2	72	25	0.25	3
8	BOTTOM ASH-4 (DISCHARGE LINE OF UNIT-4)	DS	BOTTOM ASH	2.24	1	2	80	17	0.20	3

Note: *Disturbed Sample

6 Conclusion

To ensure sustainable and environment friendly infrastructure, the suitability of bottom ash as filter material with respect to pond ash as base material has been established by NTPC for use in dyke embankment of its various plants. Out of the 10 plants explored, Bottom ash of the 9 plants has passed the required filter criterias. Based on laboratory tests, it is established that bottom ash possess the required filter ability, drainage capacity, self healing properties and does not segregate. The above have been successfully implemented and found to be safe under operating phase. During operation phase, it is also observed that cracks developed in the filter zone due to differential settlements do not stay open due to self healing property available in bottom ash. The bottom ash is a free issue material and contributes in ash utilisation. It's use reduces construction time of the dyke embankment and ensures early availability of lagoons for ash disposal to meet the operational requirement. This also saves time and cost over runs due to unexpected delays in procurement of natural sand and most importantly it is eco-friendly too.

References

Kaniraj, S.R., Gayathri, V.: Permeability and consolidation characteristics of compacted fly ash. J. Energy Eng. ASCE **130**, 18–43 (2014)

IS: 12169: Criteria for design of small embankment dams

IS: 7894: Code of practice for stability analysis of earth dams

IS: 1498: Classification and identification of soils for general engineering purposes

IS: 9429: Code of practice for drainage system for earth & rock fill dams

DOC.NO: QS-01-PEC-W-02. NTPC Guidelines for design of dykes in the ash disposal area

Long-Term Settlement (Creeping) of Soft Soils, and Ground Improvement

Heinz Brandl[✉]

Vienna University of Technology, Vienna, Austria
office@ahabrandl.at

Abstract. This keynote paper focuses on the long-term settlement (creeping) of highly compressible soils, considering also soft sludge. Long-term oedometer tests lasted up to 42 years and were performed on silty sand, (organic) clayey silt, peat and (pre-treated) sewage sludge. Secondary consolidation (creep) could be observed in all cases, lasting over many years and occurring widely linear with the logarithm of time. However, temporary acceleration may also be observed, indicating a discontinuous nature of internal deformations due to accelerated rearrangement in the fabric – mainly in very soft soils with peaty components. This long-term phase is followed by tertiary creeping with a long lasting fading out period. In both phases microcrystalline sliding occurs.

In addition to the laboratory tests results of comprehensive field measurements are summarized, showing the influence of different ground improvement methods on the creeping behaviour of highly compressible fine-grained soils (partly organic). Most data were collected from a highway junction with embankments on very soft, heterogeneous ground (locally 15 m deep and with a natural water content up to 1000%), constructed between 1972 and 1974, and monitored since. Different ground improvement methods were compared, disclosing details of primary, secondary and tertiary settlement: Deep dynamic compaction/consolidation (heavy tamping), vibro-flotation (piled embankments), temporary surcharge loading, and local combinations of the previous methods, including also vertical drains.

Keywords: Settlement · Creeping · Highly compressible soils · Oedometer tests · Heavy tamping · Deep ground improvement · Sludge

1 General

K. Terzaghi's and O.K. Fröhlich's theory of (one dimensional) consolidation refers to the dissipation of excess pore water pressure during loading of saturated soil. The time taken for the clay to consolidate depends entirely on the permeability of the laterally confined clay. These assumptions correspond to the primary consolidation in an oedometer test (widely neglecting possible rearrangements of the soil structure already in the initial phase of loading).

At the 1st International Conference on Soil Mechanics and Foundation Engineering 1936 at Harvard University, Cambridge, MA., A.S. Keverling Buisman presented a theory for creep of fine-grained soft soils. However, this (logarithmic) formula and his

© Springer Nature Switzerland AG 2020
H. Shehata et al. (Eds.): GeoMEast 2019, SUCI, pp. 125–144, 2020.
https://doi.org/10.1007/978-3-030-34184-8_9

statement that creeping of clays never ends was severely questioned, not only by K. Terzaghi (Conference Chairman), but also internationally. Meanwhile this theory has been accepted theoretically and could be widely confirmed, especially by the following test results showing low-term creep, but also a fading out tertiary creep.

2 Settlement/Creeping of Highly Compressible (Organic) Clayey Silt

Total settlement of saturated cohesive soil comprises immediate settlements (s_o – undrained, at constant volume), primary settlements (s_1 – consolidated by pore water pressure dissipation) and long-term creeping (s_2, s_3). In the field, all phases interact during transition zones, and creeping under shear stress also occurs. This leads inevitably to soil rheology comprising also cohesionless soils and other geomaterials.

In the design phase (1971–1972) of a highway junction on highly compressible soils with locally organic inclusions and peaty interlayers numerous samples were taken and investigated in the laboratory. Several of them were left in the oedometers for long-term creeping tests. The maximum observation period has been from 1971 to 2013, hence 42 years. Some results are described in the following, further investigations and in-situ influences of ground improvement measures are given in the chapter after next.

Table 1 shows the relevant data of a selected sample (A). It is an extremely soft clayey silt ($33\% < 0,002$ mm) with organic components of liquid consistency. The plasticity index of $Ip = 0.34$ is rather due to the decomposed peaty organics than to the mineralogical composition of the fines as can be seen from Table 2. The platy shape of the fines and its way of sedimentation created a special fabric and high compressibility.

Table 1. Geotechnical parameters of organic soil (sample A) and pre-treated sewage sludge (samples B, C). In brackets the years of test start.

	Sample A (1971)	Sample B (1997)	Sample C (2003)
Natural water content w_n (%)	130	168	131
Unit weight of soil particles γ_s (kN/m^3)	2.52	2.32	2.21
Void ratio e (-)	4,68	4.06	3.17
Initial dry density γ_d (kN/m^3)	0.44	0.46	0.50
Liquid limit w_l (%)	92	84	Test not possible
Plasticity limit w_p (%)	58	72	
Plasticity index I_p (%)	32	12	
Ignition loss (%)	25	35	27

In the natural state the permeability coefficient was about $k = 10^{-7}$ m/s but dropped significantly during loading. At the maximum load finally a value of about $k = 10^{-9}$ m/s was reached. These data explain, among other influence factors, the relatively quick primary consolidation and a long-lasting creeping phase.

Figures 1 and 2 show the void ratio – pressure diagram and the time-settlement curves of the particular load steps. The sample was kept under water to simulate in situ conditions and to prevent settlements by shrinking, Fig. 2 illustrates that secondary creep occurred linearly with the logarithm of time until about one year, followed by a transition period to tertiary creep which gradually leads to a fading out of the settlement. Such a behaviour coincides with site observations showing a decreasing gradient of long-term creeping plotted on semi-logarithmic scales. This coefficient was normally considered to be constant. However, even after 42 years no final value has been reached in the oedometer test, thus indicating viscous behaviour and on-going rearrangements of the soil micro-structure, due to tabular sheet silicate in connection with the loss of adhesive water, and microscopic interactions between particles and liquid. Moreover, the compression curve partly consists of segments mutually intersecting in bifurcation points which mark occasional structural collapses. This is schematically indicated in the enlarged detail within Fig. 2.

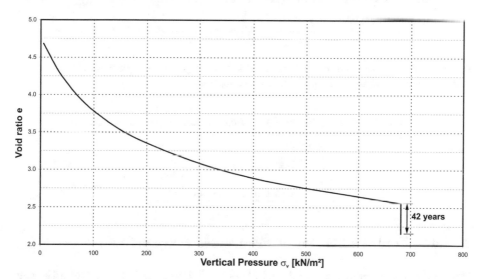

Fig. 1. Void ratio – pressure diagram (oedometer test) for organic clayey silt (sample A)

The oedometer tests were performed with incremental loading, also comprising hydraulic conductivity tests with falling height. The sample height was h = 20 mm, the diameter varied between d = 60 to 100 mm, hence providing a d:h ratio of 3 to 5 (to assess possible skin friction).

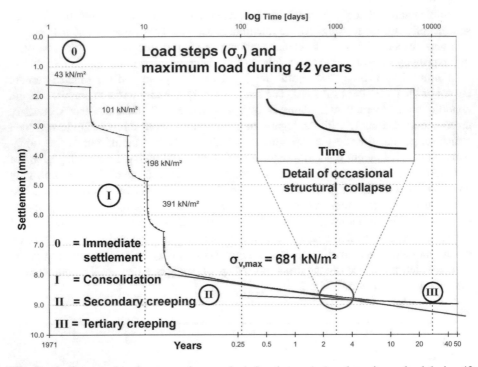

Fig. 2. Settlement – log time curve for sample A. Load steps (σ_v) and maximum load during 42 years. Also indicated are occasional structural collapses.

Table 2. Mineralogical contents of the organic clayey silt (sample A)

Mica-group	33%
Chlorite–group	16%
Quartz	40%
Feldspar (mainly plagioclase)	11%

3   Long-Term Settlement of Pre-treated Sewage Sludge

During the past decades ponds, pit landfills or surface impoundments of liquid sewage sludge have been increasingly substituted by waste deposits of pre-treated sewage sludge, unless this is not incinerated. Such landfills may reach a height of 30 m and more, thus requiring stability analyses, settlement prognoses, assessment of long-term behaviour of the liners, etc. Consequently, these aspects have become a special field of geotechnical engineering.

The settlement of pre-treated sludge is of interest not only for engineered waste disposal facilities but also for areas, where infrastructure buildings are constructed on these deposits – for instance, roads, parking places, sports grounds, light-weight structures.

Suitability tests, starting in the early 1990s disclosed that sewage sludge dewatered in a filter press and stabilized with unslaked lime can be easily deposited in all kinds of waste disposal facilities.

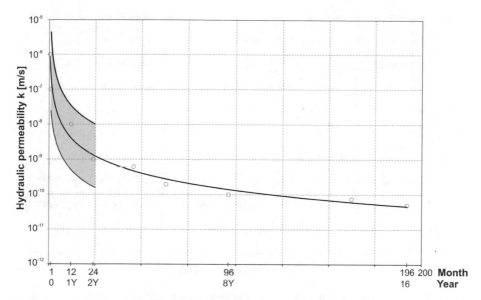

Fig. 3. Decrease of hydraulic permeability of pre-treated sludge (25–35% CaO) with time. Scatter of test series within first 2 years and example up to 16 years.

After comparative test series at the filter press and on the landfill with 20 to 45% CaO, an amount of about 31% was found to be optimal. Furthermore, 5 to 7% $FeCl_3$ was added as a flocculant. In the case of sample B 22% CaO was added (referring to the dry mass), in the case of sample C 31% CaO. When reacting with water, $Ca(OH)_2$ developed, thus creating a highly basic environment. Depending on the untreated sludge properties the hydraulic conductivity first decreased with the amount of added CaO but then increased. However, in the long-term decreasing k-values could be observed also for high CaO addition. This is rather similar to the stabilization of fine soils with lime.

Samples from the undisturbed filter-cake exhibited hydraulic permeability coefficients of only $k = 10^{-9}$ to 10^{-10} m/s. However, after field compaction of the broken filter cake these values increased to an in-situ permeability of about $k = 10^{-7}$ to 10^{-8} m/s, though the dry density was only $\rho_d = 0.45$ to 0.55 g/cm^3 (water content usually about w = 130%). In the long-term in-situ values down to $k = 10^{-9}$–10^{-10} m/s were measured, depending on the amount of added lime.

The k-value decreased with time due to mechanical, chemo-physical and biological long-term reactions. In the laboratory, values of $k = 5.10^{-7}$ to 10^{-9} m/s were measured within six months of curing, depending on vertical load and CaO additives. Figure 3 shows an example of long-term tests (running 16 years) together with the scatter of several test series with 25 to 35% CaO within the first two years at a vertical pressure of

250 kN/m^2. Hence, pre-treated stabilized sludge can be thoroughly considered as secondary barrier material within the sealing system of a waste deposit. However, compaction in layers is essential.

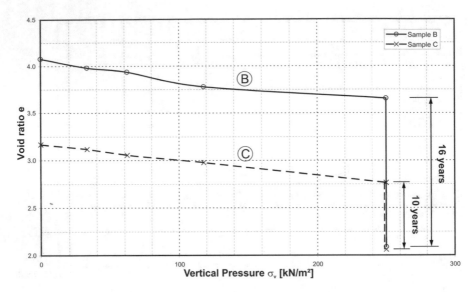

Fig. 4. Void ratio – pressure diagram (oedometer tests) for pre-treated sewage sludges (samples B, C).

In order to investigate the long-term behaviour of pre-treated sewage sludge and to find analogies between sludge and soil behaviour several samples were taken. The focus has been on creeping because this has the largest influence on the long-term behaviour of the surface liner of a waste deposit with regard to (differential) settlements.

In the following two examples are selected including long-term oedometer tests running from 1997 and 2003 respectively, until 2013. The samples were always under water and exposed to a constant temperature of 20 °C (±1 °C).

Table 1 summarizes the most important geotechnical parameters. The particle size distribution shows "clayey sandy silt" with rather uniform mineralogical contents: Mainly calcite due to the CaO additives, further quartz and some feldspar and layer silicates. Chemical investigations found some concentration of zinc, copper and lead. The material exhibited liquid consistency and zero to low plasticity. The permeability factor was about $k = 10^{-6}$ m/s at the beginning of the compression (oedometer) test under the load step of $p = 30 \text{ kN/m}^2$ and decreased to about $k = 10^{-10}$ m/s after 15 years under $p = 250 \text{ kN/m}^2$. The stress-void ratio diagrams show compression curves similar to natural soils (Fig. 4) but less curved and with strong long-term compression under the maximum load.

Figures 5 and 6 show the settlement - time diagrams in semi-logarithmic scale. They illustrate that within the first weeks the settlements were rather small, even under the maximum load step. Then they increased significantly, similar to very soft soils.

After one year this intensive consolidation was nearly abruptly followed by creeping comprising mechanical, chemo-physical and anaerobe biological reactions. The latter might be the main reason that creeping of sample B does not occur linearly with logarithm of time but in a slightly convex curve (Fig. 5).

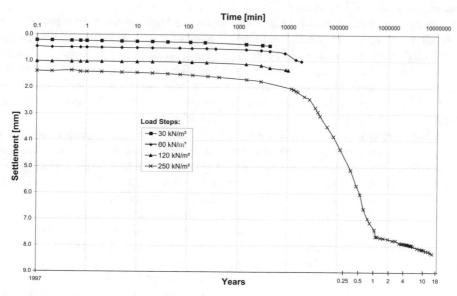

Fig. 5. Settlement – log time curves for sample B and increasing load steps (max. 16 years)

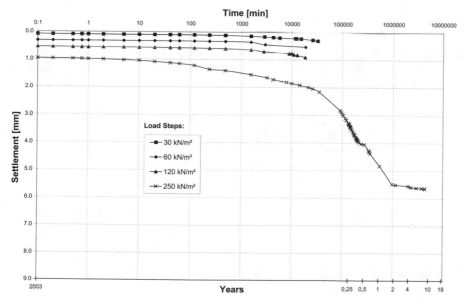

Fig. 6. Settlement – log time curves for sample C and increasing load steps (max. 10 years)

It is noticeable that the hydraulic permeability decreased most in the first year – corresponding to the settlement curve. Long-term pore clogging is influenced by particle rearrangements, lime reactions and possibly biological activities. The sample investigated since the year 2003 was obviously lime-saturated: Repeated hydraulic permeability tests with the oedometer caused – mainly in the first phase – some washing out of calcitic particles.

Creeping continued until the end of the oedometer tests, i.e. up to 16 years without coming to an end. This clearly indicated a long-term rearrangement of the sludge structure despite the hardening effect of added lime. Similar behaviour could be found for inorganic clayey silt and silty clay stabilized with lime, when cured under water-saturated conditions. However, creeping of such soils faded out at least within ten years. In both cases (sludge and soil) the creeping value (i.e. the gradient of the settlement line) dropped with increasing amount of added lime.

Chemo-physical and anaerobe biological reactions of sludge explain a long-term creeping of sewage sludge, which sometimes differs from natural soil or peat. Though the absolute values are small, the settlement - log t correlation may be not a straight line but slightly curved downward – depending on organisms and chemistry (e.g. Fig. 5). Nevertheless, settlements in the oedometer cannot proceed indefinitely.

Unloading of the long-term oedometer tests showed only small swelling. This is due to the high amount of added CaO and the non-active mineralogical contents.

The hitherto field observations confirmed the results of laboratory and in-situ tests. Primary consolidation of the waste deposit occurred already during the several years lasting landfilling process, and long-term creeping is no problem for the sealing cover. It is smaller than under saturated laboratory conditions because of gradual carbonatisation of the material.

4   Theoretical Aspects of Creeping

Regarding the theoretical aspects of vertical and horizontal consolidation, and secondary compression/creeping several hypotheses are existing. One of them is indicated below (Thurner et al. 2019):

$$S = \underbrace{\frac{C_C/C_S}{(1+e_0)} H \log\frac{\sigma'_f}{\sigma'_{vo}}}_{\substack{\text{primary settlement} \\ \text{(consolidation)}}} + \underbrace{\frac{C_\alpha}{(1+e_{100})} H_{100} \log\frac{t}{t_p}}_{\substack{\text{secondary settlement} \\ \text{(creeping)}}} \qquad (1)$$

McCarthy (2002) calculates the amount of secondary compression ΔH_{sc} as:

$$\Delta H_{sc} = H_0[C_\alpha/(1+e_0)]\Delta\log t \qquad (2)$$

In analogy to consolidation Eq. (2) results also from $\Delta e = C_\alpha \cdot \Delta\log t$.

Holtz and Kovac (1981) and Das (1998) differ from (2) by considering the void ratio at the end of the consolidation or primary compression e_p:

$$\Delta H_{sc} = H_0 \left[C_\alpha / \left(1 + e_p \right) \right] \Delta \log t \tag{3}$$

Das (1998) acknowledges the two alternate definitions for the coefficient of secondary compression and gives the expression that relates the two (C_α in terms of coid ratio. C_α' in terms of strain):

$$C_\alpha' = C_\alpha / \left(1 + e_p \right) \tag{4}$$

Budhu (2000) agreeing with Eq. (3), is a little more precise in the time term:

$$\Delta H_{sc} = H_0 \left[C_\alpha / \left(1 + e_p \right) \right] \Delta \log \left(t / t_p \right) \tag{5}$$

where t_p is the time when the primary compression ends.

Equation (1) assumes that creep starts after primary consolidation, whereas other hypotheses assume that creep starts from the beginning, i.e. from e_0. Actually there is a transition zone, its period depends on the soil parameters and external loads (magnitude and time of load steps). Consequently, an optimisation of settlement analyses is most efficient if using calibrated parameters from (large scale) test fields. Hence, C_c should be measured during the primary stage regardless if it contains creep or not, and C_α should be measured during the secondary stage, regardless if this started earlier or not.

Contrary to K. Terzaghi's conventional theory the coefficient of primary consolidation is not a constant parameter. However, the creeping coefficient is widely constant during seconderary and tertiary creep.

5   Influence of Ground Improvement on Long-Term Settlements

Between 1972 and 1974 a large highway interchange was constructed on highly compressible heterogeneous ground (Tauernautobahn, Austria). It comprised embankments up to 8 m height, max. 2.5 m deep excavations mostly in peat and 8 ridges. The following ground improvement methods were applied (Fig. 7; details see Brandl 2006):

- Deep dynamic compaction/consolidation (heavy tamping),
- Vibroflotation,
- Temporary surcharge loading,
- Local combinations of the previous methods.

Fig. 7. Vibroflotation and early heavy temping (22.5 tons from 25 m) in Austria (1972). Highly compressible, extremely heterogeneous soil.

Deep dynamic compaction by heavy tamping has been used in Austria and Germany since the 1930s, but was first limited to granular materials, drop weights of about 10 tons and drop heights of about 10 m. Significant development started at this construction site in 1972/73 with 20 to 25 tons falling from heights up to 22.5 m to improve soft or loose soils respectively and peat to a depth of about 14 m. This required special crawler cranes and in advance fill layers as working platform.

An impact "consolidation" of more or less water saturated (organic) clayey silts and peat was considered "impossible" at that time as being completely contradictory to K. Terzaghi's and O.K. Fröhlich's consolidation theory. Fortunately, the owner (Austrian Federal Ministry) could be convinced to allow an increased geotechnical risk in the frame of research and development, and to reduce costs and construction time. Intensive site observations and measurements disclosed that the excessive impacts created from heavy tamping on the soil caused particle rearrangements, local soil liquefaction and steep shear surfaces where vertical drainage up to the ground surface occurred (like artesian water). This behaviour was favoured by (micro)gas bubbles in the soft soil: 100% water saturation is hardly measured in practice, even in inorganic fine-grained soils below groundwater. This could be observed on numerous construction sites.

The thickness of the highly compressible and heterogeneous layers varied between 3 to 16 m, comprising peat, clayey to sandy silt, silty sand (locally with gravel), and finally sandy gravel. Organic interlayers were found down to 15 m below original ground. The groundwater level depended strongly on weather and season with a mean value of approximately 2 m below surface (Fig. 8).

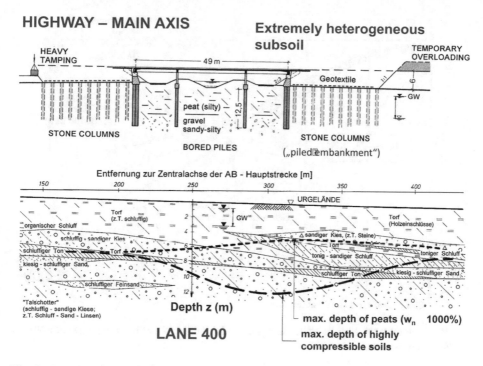

Fig. 8. Longitudinal section along lane 400 of a highway junction on highly compressible, extremely heterogenous ground.

Laboratory tests and in-situ measurements provided compression moduli down to $E_s = 0.2$ MN/m^2 and a natural water content up to about $w_n = 1000\%$. The saturation degree varied between 75 to nearly 100%, clearly increasing below groundwater table and with depth. The liquid limit lay between $w_L = 20$ to 600%, the plasticity index between $I_p = 0$ to 250%. These extremely poor ground conditions led to settlements up to about 5 meters already during the construction process. Further details can be obtained from (Brandl 2006).

Oedometer tests on organic soils showed a significant tendency to creeping. According to Fig. 9 a creeping coefficient for secondary settlements was derived. The transition from primary to secondary settlement is indicated in Fig. 9 by an idealized line, but actually occurred within a longer period. Figure 10 shows that the creeping coefficient varied within a very wide range. Several oedometer tests ran over a period of 20 years and one test up to 42 years (from 1971 to 2013). These long-term investigations have disclosed that secondary and tertiary creep may continue extremely long if the fine-grained soil has a high void ratio and organic components. The mineralogical composition of the fines is also of influence.

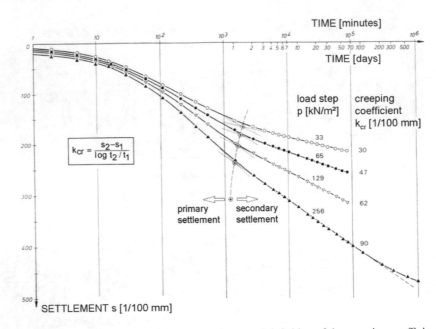

Fig. 9. Time-settlement curves of decomposed peat and definition of the creeping coefficient k_{cr} (derived from oedometer tests).

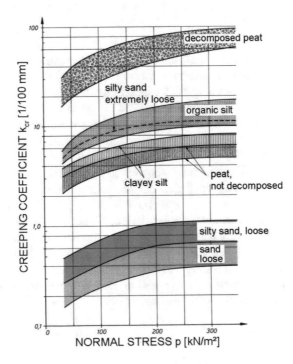

Fig. 10. Creeping coefficient k_{cr} for several soils describing secondary settlement (creeping).

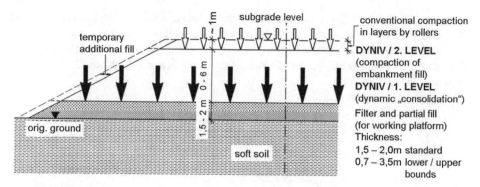

Fig. 11. Standard procedure of heavy tamping at a highway interchange performed in the years 1972/1973

The scheme of Fig. 11 illustrates the compaction procedure typically applied for the embankments, whereas sections below original ground surface required a partial soil exchange before heavy tamping. Due to the heterogeneous subsoil and varying embankment heights or cut depths respectively the required compaction energy varied in a wide range with a maximum of approximately $E = 2500$ tm/m^2. Deep compaction control was performed mainly by comparing pressuremeter values before and after heavy tamping. Figure 12 shows an example illustrating the influence depth of heavy tamping and the effects of the embankment weight and time. The influence depth of heavy tamping varied between 8 to 14 m depending on particular soil properties and energy input.

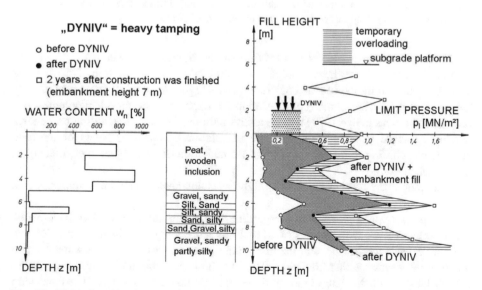

Fig. 12. Example of in-situ pressuremeter tests before and after heavy tamping, and two years after the embankment had been constructed

Figure 13 presents the settlements of an interchange section where seven series of heavy tamping and a temporary surcharge load on the embankment were applied. The final road pavement was installed approximately 16 years after opening of the highway. The secondary settlements within this period did not affect the highway traffic as they occurred rather uniformly.

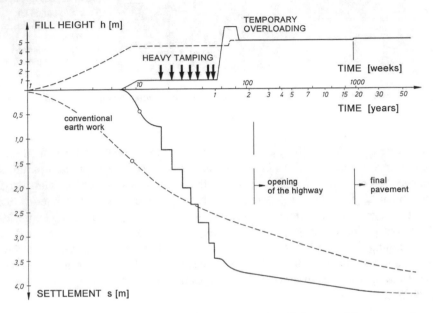

Fig. 13. Time-settlement curve of an embankment section. Influence of heavy tamping and temporary surcharge load on the level of the embankment crown

Figure 14 however, shows that section of the interchange, where the maximum settlement occurred after heavy tamping and embankment construction. This required periodical relevelling despite the construction of a higher level of the pavement already before opening for the traffic (compensation for expected long-term settlements). But only the first measure (in 1977) was an additional one; the other relevelling procedures where performed in connection with the installation of the final layers of the road pavement according to the original design (remediation of wearing courses, placing of drain asphalt etc.).

The long-term behaviour of this highway interchange may be summarized as follows:

The project was a pioneer work regarding heavy tamping and piled embankments. Previous experience with weights of 20 to 25 tons dropping from heights up to 22.5 m did not yet exist worldwide, and the fine-grained, organic ground with a water content up to 1000% was another challenge. Despite these unfavourable conditions a satisfactory long-term behaviour of the entire interchange could be achieved. The maximum total settlement (including anticipated deformations by heavy tamping and temporary surcharge loading of the embankments) was approximately 5 m which occurred mainly during the construction period. Local re-levelling of the primary (provisional) road surface on the basis of the contractor's quality guarantee was necessary only once, namely 2.5 years after opening of the highway. This measure was limited to some sections of heavy tamping only.

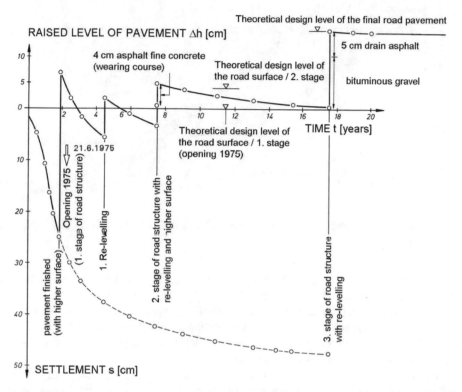

Fig. 14. Time-displacement curves (related to the design level of the road surface) involving periodical re-levelling and installation of additional surfacing layers to achieve sufficient driving comfort (Section 10).

Long-term creeping after highway opening varied between 5 to 20 cm and occurred rather uniformly. The piled embankments resting on stone columns with compound body cover (geosynthetics, cement stabilization, crushed rock) behaved even better than the sections with heavy tamping. However, the ground properties were somewhat better there. Temporary surcharge loading of the embankment proved to be also very successful, especially in connection with previous heavy tamping.

According to Austrian highway guidelines and codes the definite surfacing of the road pavement was placed approximately 4.5 years after opening of the highway (2nd stage of road structure). The final surfacing, 16 years after opening, involved the placement of a new road structure with a more traffic resistant wearing course above the old structure. This remediation was required primarily because of the long-term degradation of the road pavement (deep traffic ruttings etc.) due to heavy traffic. The influence of differential settlements was negligible. However, both road surfacing measures (4.5 and 16 years after highway opening) involved also a re-levelling.

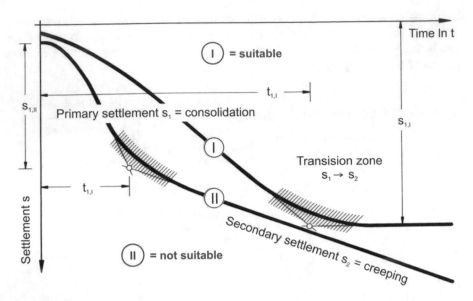

Fig. 15. Influence of consolidation and creeping behaviour of soils on the efficiency of vertical drains (schematic)

To sum up, the long-term behaviour of this highway interchange has been very satisfactory for about 40 years now. The design speed of v = 150 km/h could be maintained during the entire period. The settlement prognoses based on laboratory and field tests, on analytical calculations, on empirical parameters and experience have been in good accordance with the measured values. Long-term creeping is still going on but negligible for traffic comfort and maintenance.

Comprehensive field observations have disclosed how method and quality of deep soil improvement influence primary consolidation and creeping of soils. Consequently, if a ground tends to strong creeping (observed in laboratory tests), soil improvement technologies have to be properly selected or adapted, resp. For instance, vertical drains accelerate only pore water dissipation during primary consolidation, but do not improve creeping behaviour.

Prediction of primary and secondary settlements was important for constructing a temporarily higher level of subgrade and asphalt surface of the highway junction running on the embankments on highly compressible ground ("compensation fill").

Several comparative tests showed that Atterberg limits or activity index, resp. are not sufficient as a criterion for creep assessment, because soil creep depends on numerous factors: Grain size distribution, mineral optical composition, moisture content, permeability, density, fabrics structural strength, viscosity, and external factors lead to an extremely complex process.

6 Creeping and Efficiency of Vertical Drains or Geosynthetics

The tendency of creeping has a great influence on the efficiency of vertical drains and on the long-term performance of embankments on very soft soil. The scheme of Fig. 15 indicates the difference between a soil with clearly dominating consolidation and negligible creeping on one hand, and less consolidation but excessive long-term creeping on the other hand. Also indicated is an idealised transition zone. Consequently, vertical drains have a clearly higher efficiency in case I and are hardly suitable for excessively creeping soils (case II).

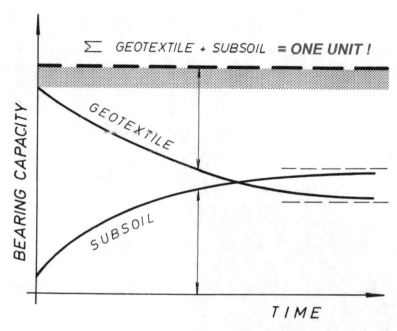

Fig. 16. Long-term behaviour of basal reinforced (low) embankments on very soft soil due to creeping of geosynthetic and consolidation/compression of soil (schematic)

Especially in the early phase of geotextile application in soil mechanics there were severe discussions and debates about creeping took place (Brandl 1986). A main topic was the long-term bearing behaviour of low embankments on very soft soil. Field observations since the early 1970s have confirmed that the interacting system geosynthetics plus subsoil should be considered as one unit with a mostly rather constant long-term bearing capacity for roads and railways. The scheme of Fig. 16 indicates a certain time dependent reduction for geotextile/geosynthetics (material creeping) but an increasing bearing capacity of the soil due to consolidation and compression.

Parallel to a severe scepticism against an innovative application of geotextiles in the early 1970s over-optimistic groups sometimes even ignored basic soil mechanics. A certain overestimation of the efficiency of geosynthetics could be observed, just as a "fundamentalist" rejection. Figure 17 shows an example of special ground failures of embankments on extremely soft soils. Too quick filling of the embankment sometimes caused failure modes completely different from the "classical" one, depending on the stress-strain behaviour of (very) flexible geotextiles. Consequently, accelerating the earthworks (fill) on highly compressible, fine grained soil in order to reduce the construction time requires a detailed investigation of soil-geosynthetics interactions.

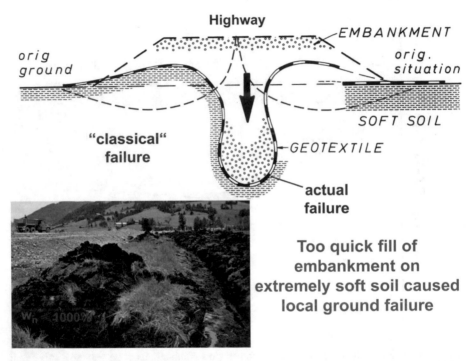

Fig. 17. Ground failures in extremely soft organic soil (natural water content up to 1000%).

7 Dynamic Soil Improvement

The project of 1972 and its long-term observation described in this paper had pioneering character. Heavy tamping of deep reaching clayey, silts and fine-grained organic layers with interlayers of peat required an amount of energy, never exceeded since. Meanwhile equipment and capacity of dynamic soil improvement have improved significantly. This is mainly based on electronic recording the interaction between soil and equipment with continuous interpretation and simultaneous machinery reaction. Figure 18 gives a schematic overview of the present state of the art with the maximum influence depth that can be reached economically.

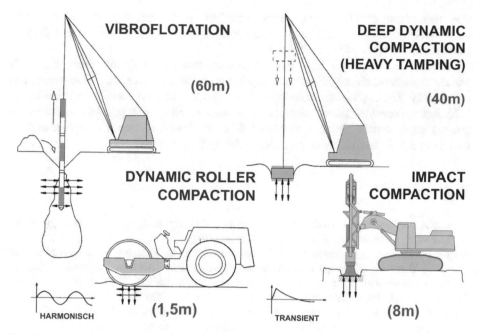

Fig. 18. Dynamic soil improvement: technologies and economically maximum influence depth.

8 Conclusions

All three-phase systems containing particles, liquids and gas exhibit creep under compressive stress. Secondary creeping of clayey soft soils mostly occurs linear with the logarithm of time. However, temporary increase may also be observed, indicating a discontinuous nature of internal deformations due to accelerated rearrangement in the fabric – mainly in soils with peaty components. A micro-mechanical explanation is ductile sliding between mineral crystals followed by repeated structural ruptures.

Long-term oedometer tests on soils and pre-treated sewage sludge have revealed several similarities between natural and artificial fine materials of high compressibility. The tests ran up to 42 years and showed a gradual transition from secondary to tertiary creep for organic clayey silts after about one year. During tertiary creep the gradient, plotted on semi-logarithmic scale, gradually decreased. This could be found also for inorganic clays under site conditions, where the gradient may eventually approach zero.

In pre-treated sewage sludge the transition from primary to secondary consolidation is more significant than in soils. No fading out tertiary creep could be observed in the semi-log diagrams of oedometer tests. This could be explained by chemo-physical and anaerobe biological long-term reactions in this material.

Several other comparative tests have confirmed that organic soils show pronounced secondary/tertiary creeping, and that creeping also depends on the mineralogical composition and the arrangement (microscopic structure) of the fines, and not only on grain size distribution, initial porosity, consistency, etc. Consequently, prognoses of creeping derived from oedometer tests (with site-specific data) are still more reliable than those from exclusively numerical modelling (with data from the literature) or uncertain correlations. Additionally, investigations based on geochemistry and electron microscope analyses might be helpful.

Finally, the long-term tests have disclosed, that K. Buisman's creep theory is widely consistent, although tertiary creep has to be added and settlements cannot run indefinitely. The application of oedometric results for the prediction of creeping in the field has to consider lateral displacements and shearing and possible measures of ground improvement. Three-dimensional field conditions accelerate creeping in relation to one dimensional oedometer tests (with stiff side walls).

References

Brandl, H.: Research and development in geosynthetics engineering. In: Proceedings of 3rd International Conference of IGS, vol. 5, Vienna (1986)

Brandl, H.: Ground improvement and earthwork innovations for transportation infrastructure. In: Active Geotechnical Design in Infrastructure Development. XIII Danube-European Conference on Geotechnical Engineering, CIP-Ljubljana, vol. 1, pp. 217–232 (2006)

Budhu, M.: Soil Mechanics & Foundations. Wiley, New York (2000)

Buisman, A.S.K.: Results of long duration settlement observations. In: Proceedings of the 1st International Conference of the ISSMFE (International Society for Soil Mechanics and Foundation Engineering), Cambridge, vol. 1, pp. 103–106 (1936)

Das, B.M.: Principles of Geotechnical Engineering, 4th edn. PWS Publishing Company, Boston (1998)

Havel, F.: Creep in soft soils. Doctoral Thesis, Norwegian University of Science and Technology, Trondheim (2004)

Holtz, R.D., Kovacs, W.D.: An Introduction to Geotechnical Engineering. Prentice Hall, Englewood Cliffs (1981)

Sivakugan, N., Das, B.M.: Geotechnical Engineering. A Practical Problem Solving Approach. J. Ross Publishing, Fort Lauderdale (2010)

Terzaghi, K., Fröhlich, O.K.: Theorie der Setzung von Tonschichten. Franz Deuticke, Leipzig – Wien (1936)

Thurner, R., et al.: Vertikaldrainagen – Detaillierte Bemessung, Vergleich mit Messdaten und Anwendung im großen Maßstab. In: 12th Austrian Geotechnical Conference, Vienna (2019)

Equations to Correct SPT-N Values Obtained Using Non-standard Hammer Weight and Drop Height – Part III

Khaled R. Khater[1]([⊠]) and Mohamed A. Baset[2]

[1] Civil Engineering Department, Fayoum University, Fayoum, Egypt
krk26@fayoum.edu.eg
[2] Fayoum, Egypt

Abstract. The authors previously published two papers to study the influence of using non-standard hammer-weight (W_i) on the obtained blow counts, (N_i) during the SPT test. The current study is an extension to the past work; here the hammer drop height (H_i) has been changed to investigate its effect on the obtained blow counts, (N_i). Then, the multiple action of changing non-standard energy ($W_i.H_i$) is also investigated. The hammer weight, its drop height and the machine efficiency ($\mathcal{e}_i$) present the three main variables during the study.

The study methodology is experimental work and the method is physical model simulates the behavior of SPT-Rig. It has been designed and manufactured in house. The model is capable to change the hammer weight, its drop height and the overall machine efficiency. Five hammer weights, five drop heights and six different efficiencies have been changed all together; individually as well as in groups. The hammer weight ratios (W_i/W_s) and drop height ratios (H_i/H_s) each has been changed to cover the spans of 0.2, 0.4, 0.6, 0.8 and 1.0. Furthermore, six efficiencies ($\mathcal{e}_i$) have been changed with every change of the above ratios to cover the scope of, $\mathcal{e}$ = 35%, 50%, 60%, 70%, 85% and 100%. Reconstituted well graded clean silicate dense sand *(SW)* is used for modeling the soil.

Novel procedure has been advised as well as empirical equations have been proposed to re-adjust the incorrect blow counts (N_i) obtained using non-standard SPT parameters or uses machines of a low efficiency. Also, a simplified method fast and costless has been suggested to sort out the problem of machines owing low efficiencies. Finally, some modifications to well-known formula are advised.

1 Introduction

Standard penetration test, SPT is part of every single soil investigation program essentially where the obtained soil samples are disturbed and unable to provide mechanical properties. If the only advantage of SPT is predicting in-situ density of a stratum, this is enough to be acknowledged. Not to mention, its advantages when it comes to predict strength and deformation parameters, through correlations.

Basically, SPT-test is a spoon driven into soil stratum, the cumulative effective blow counts N to get through this stratum a feet deep is stated. Then N corrected,

© Springer Nature Switzerland AG 2020
H. Shehata et al. (Eds.): GeoMEast 2019, SUCI, pp. 145–165, 2020.
https://doi.org/10.1007/978-3-030-34184-8_10

normalizes and correlated to soil properties. The test is first announced by Terzaghi (1947), and then every geotechnical code spread a clause for SPT-Standards.

This paper is not focused on correlations between blow counts N and soil properties. Also, it is not paying attention to whichever type of N corrections, except three major corrections. They are, hammer weight, hammer drop height and the rig overall efficiency. Previous experience shows evidences of non-standard hammer weight used as well as insufficient hammer drop height. Also, due to lake of maintenance, the SPT-rigs overall efficiency decreases. The later one by itself misleads N-results, let alone if it goes with any of the other two evidences. The research question was, if one of the three defects already occurred could it be corrected afterward?

To answer this, a series of four research papers has been planned to investigate the above mentioned three imperfections. Non-standard hammer weight and rig overall efficiency had been investigated throughout three papers already published. The current paper is the fourth. It investigates the case of non-standard hammer drop height. The four researches are experimental performed based on physical model simulates the SPT-rig. The physical model is filled with reconstituted well graded dry dense sand. This paper hints: topic is closed based on plan, but the criticism of the four integrated papers activates brain storming for refinements.

This manuscript consists of five main titles. Its starts with a brief literature review for works published by others. It is followed by summary and main findings of our previously published papers in topic. Next the physical model and the used reconstituted soil are presented. The case of non-standard hammer drop height is the focus of this paper, accordingly discussed in details. Lastly, the main findings of the four papers are summed up to complete the panorama and scope of this topic.

2 Literature Review

The literature review consists of four main headings. It starts with the concept of energy ratio normalization, $(ER)_{60}$. Then, samples of equations suggested by pioneers, scientists and committees have been discussed. These equations were developed to correct the measured blows count N_m obtained using non-standard test parameters. The authors of the current paper developed additional equations to correct N_m obtained using non-standard hammer weight, throughout three published papers. Those papers recapped and presented as part of our literature review. This writing-up technique connects our previous work to the current. The results of this study are equations to correct N_m measured from non-standard drop height. So, the four papers integrated and formed a practical correction system. The literature review has been ended-up by the three magic words gap, hypothesis and research question.

2.1 Normalizations of N_m for $(N_1)_{60}$

Prior the step of energy ratio normalization, $(ER)_{60}$ The normalization of the measured blow counts N_m is discussed. As an example, Fang (1999) recaps the normalization practice to adjust the measured blow counts N_m to be $(N_1)_{60}$. Also, Tokimatsu and Seed (1987), added to the topic the effect of overburden pressure adjustment C_N and the liner adjustment factor C_S.

2.1.1   Energy Ratio Normalization

The energy ratio ER is the ratio of the measured hammer efficiency, E_m to theoretical free-fall hammer efficiency, E_{th}. The Normalized energy ratio $(ER)_{60}$ is the ratio of the measured hammer efficiency, E_m to 60% E_{th}.

$$E_{th} = \frac{1}{2}mv^2 = \frac{1}{2}\frac{w}{g}v^2 = \frac{1}{2}\frac{w}{g}\left(\sqrt{2gh}\right)^2 = wh \tag{1}$$

Where m is the hammer mass, v is the fallen velocity and g is the acceleration. The In-Situ applied energy that denoted by E_m has been measured and practically found to be, in average about 50% of E_{th}. It presents the efficiency of hammer and it's over all components. In that case, the normalized energy ratio $(ER)_{60}$ is as follows, e.g.:

$$(ER)_{60} = \frac{E_m}{E_{th}} = \frac{50\%E_{th}}{60\%E_{th}} = 0.83 \tag{2}$$

Skempton (1986) is first one suggested standardizing results of hammers having varying efficiencies to be as 60% E_{th}. The typical practical ranges for $(ER)_{60}$ are 0.5:1.0 Donut, 0.7:1.2 Safety and 0.8:1.5 Automatic hammer. Rami (2013) evaluated the energy efficiency of hammers used in Egypt. The results show normalized energy ratio $(ER)_{60}$ of 0.82 and 1.0 for Donut and Safety hammers, respectively.

2.1.2   Significance of $(N1)_{60}$ Normalization

FHWA (2006) recommended to adjust the measured N_m to be $(N_1)_{60}$. Figure 1, shows two SPT-tests results conducted by donut and safety hammers inside same borehole. The energy ratio ER found to be 45% for Donut and 60% for Safety hammer. Graph (a) shows individual trends of N_m values for both hammers. A consistent profile is obtained as shown in graph (b) once data were normalized to be $(ER)_{60}$.

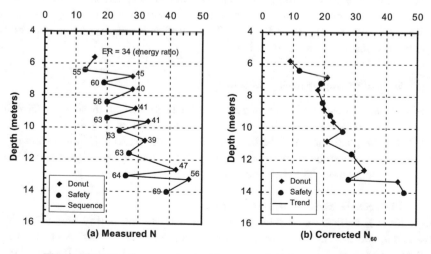

Fig. 1. SPT N-values vs. depth: (a) measured & (b) corrected to 60%

2.2   Corrections Suggested by Other Researchers

The hammer weight W, drop height H, spoon penetration depth L and sampler diameters D are the constants that possibly may deviate of standard values and need corrections. The following are two effective and trusted correction examples:

Burmister (1948) proposed Eq. 3 based on principal of super position for correcting the measured N_m to be the corrected N_{cr}, if all or one of the up-mentioned parameters deviated. The formula assumed the variation behavior of W and/or H is linear.

$$N_{cr} = N_m \frac{(W \ Ibs)(H \ in)}{(140 \ Ibs)(30 \ in)} \frac{\left[(2.0 \ in)^2 - (1.375 \ in)^2\right]}{D_o^2 - D_i^2} \tag{3}$$

N_{cr} is the corrected N_m after using non-standard parameters. W, H, D_o and D_i are nonstandard hammer weight, drop height and outer-inner spoon diameters.

Lacroix and Horn (1973) simplified and improved Eq. 3 by skipping effect of sampler inner diameter D_i and added effect of spoon penetration depth deviation, L_i, Eq. (4). The behavior of Eq. (4) is the same as Eq. (3), i.e. the variation behavior with respect to W and/or H is linear.

$$N_{cr} = N_m \cdot \left(\frac{2 \ in}{D_i}\right)^2 \times \frac{12 \ in}{L_i} \times \frac{W_i}{140 \ Ib} \times \frac{H_i}{30 \ in} = \frac{2 \ N_m \ W_i \ H_i}{175 \ D_i^2 L_i} \tag{4}$$

Lamb (2000) replaced the 140 lbs hammer by a custom-made one of 100 lbs, and then the energy transferred was measured. The efficiency ratio has been reduced from 90% to 66%, i.e. reduced by 27% while W has been reduced by 29%, it shows no wide difference, i.e. linear behavior is acceptable approximation.

Youd et al. (2008) performed a full scale study to investigate the hammer energy ratio ER versus drop height. A 127 mm long sleeve was placed in the hammer mechanism to reduce the drop height from 762 mm to 584. The sleeves were inserted with lengths 177 mm, 127 mm, 50 mm and no sleeve. Resulting drop height were 762 mm, 711 mm, 635 mm and 584 mm. The main result was the energy transfer ratio ER increases linearly with the drop height.

2.3   Corrections Suggested by Authors

Here-in the main results of three papers previously published by authors have been recapped. The detailed methodology, method, presentation and analysis of results intentionally skipped and could be found in the original papers. This is a summery its purpose is connecting the running paper with our previous to complete the brainwave. The three papers below have been arranged chronologically.

2.3.1   Pilot Study and Premature Model

Khater (2016), carried out a pilot study to investigate the ability and reliability of constructing and then using physical model to simulate the SPT task. An earliest small physical model, first of its kind, has been designed, manufactured and then used owing a single energy ratio; ER equal to 100%, i.e. no energy loss, theoretically speaking. Next, several hammer weights have been changed and used. The corresponding blow counts reported. After a simple but representative and accurate analysis, Eq. 5 has been suggested to correct the misleading number of blows counted from the use of non-standard hammer weight.

$$N_{Std} = \eta_K \left(\frac{W_f}{W_{Std}} \right) N_f \qquad (5)$$

Where W_f is the non-standard hammer weight, W_{Std} is the standard hammer weight, N_f is the measured number of blows prior the correction, N_{Std} is the desired corrected number of blows after the use of non-standard hammer. The "η_K" is a factor its value depends on the overall system efficiency, energy transfer ratio ER. The used physical model was manufactured smooth, theoretically ER equal to 100% and hence the value of η_K is unity. That is to say, when the used hammer weight is the standard one, i.e. $W_f = W_{Std}$ and the efficiency is 100%, definitely $N_f = N_{Std}$.

This pilot study together with Eq. 5 believes that the relationship between W's and N's are linear. The line should pass by the origin and its slope with the horizontal axis varies with the variation of the model overall efficiency. The maximum value of the line slope is 1:1, that is to say, inclined by 45° where ER = 100%. Accordingly, correcting the measured blow counts N_m is possible if family of equations have been developed as a function of different values of energy ratio ER, efficiencies. This could not be done without the aid of advanced physical model. Accordingly, developing mature physical model, and choosing its parameters demanded the theme of the next research works.

2.3.2   Mature Model and Introductory Study

Khater et al. (2019a, b) Part I, published preliminarily results of an experimental study presents the extension of the above mentioned pilot study. Advanced physical model has been designed and manufactured to simulate the behavior of the SPT. The model has the ability of changing hammer weight, drop height, and efficiency levels. Starting from perfectly smooth ER = 100% and up to extremely rough ER = 35%. Throughout the published paper the results of only two efficiencies were displayed they are the efficiency borders, i.e. ER = 100% and 35%. Five different hammer weights have changed to generate the qualitative data analyzed right through. Reconstituted well graded dry dense sand SW characterized the used soil. The mature physical model is the same one used all over this series of research work. Accordingly, its full details will be illustrated and discussed later on at the methodology part of the running paper. However, Fig. 2 shows the strategy which will be used in the recent and future results presentation. Also, the preliminarily results of the two efficiencies present samples of the research product which will extended later on through a series of published researches. Furthermore, the used model proved itself as a quiet good method to handle this style of problems.

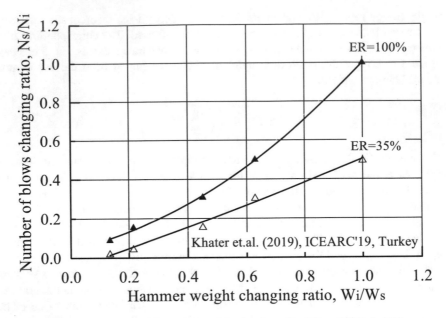

Fig. 2. Hammer weight ratios vs. blow cunts ratios, ER = 100% & 35%

Throughout the rest of this work, the terminology efficiency is the same as the energy ratio ER definition. To recap, ER is the ratio of the measured hammer efficiency E_m to the theoretical free-fall hammer efficiency E_{th}. Hence, ER = 100% means no energy losses, as an example, while ER = 35% presents rough system with energy losses up to 65% of the total applied theoretical energy.

The x-axis presents the weight changing ratios while y-axis presents the corresponding number of blows changing ratios. Each case of ER is plotted and presented as the best fit. To generalize the use of curves, absolute values were avoided and normalization system has been adopted. The largest weight and its corresponding number of blows have been considered as reference values. Accordingly, x-axis co-ordinates are W_i/W_s, y-axis co-ordinates are N_s/N_i where "i" increases from 1 to 5. W_s is the largest used weight, logically N_s is the lowest blow counts and always corresponding to W_s. This normalization system makes ranges the ratios from 0.0 to 1.0, i.e. practical and logical range.

Figure 2, displays two curves, they present efficiencies ER = 100% and 35%. It is obvious as the ratio W_i/W_s decreases, N_s/N_i decreases too. This says if the used hammer is lighter than standard the measured N_i is over estimated. Hence misleading, hammers weight wise. Same observation extends for efficiencies, as ER decreases; N_i increases, efficiency wise. Last observation must be taken seriously.

2.3.3   Mature Model and Advanced Study

Khater et al. (2019a, b) Part II published the final results as an extension to Part I to finalize the use of non-standard hammer weight issue. The same physical model as well as the same soil used in Part I, have been used too in Part II. The physical mature model manufactured with capability of changing hammer weight as well as efficiency levels. Five hammer weights per each one of six different efficiency ratios ER have been investigated and analyzed. Numerically they are efficiency ratio, ER = 100%, 85%, 70%, 60%, 50% and 35%. The hammer weight ratios W_i/W_S are 0.13, 0.22, 0.45, 0.63 and 1.0, the results are shown in Fig. 3. The system of results presentation and axes normalization used in the Fig. 2 has been adopted here-in.

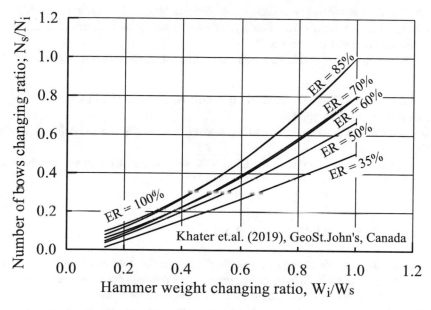

Fig. 3. Hammer weight ratios vs. blow cunts ratios, ER = 100%:35%

As shown in Fig. 4, for same W_i/W_s if ER decreases N_s/N_i decreases, i.e. N_i increased. Also, at W_i/W_s equal to 1.0, i.e. standard hammer is used, if ER decreases N_s/N_i decreases. In other words, in the above mentioned examples N_i overestimated. This is over looked by many designers and this is non-seen damage. The relationship between W_i/W_s and N_s/N_i is linear at low values of ER and lightly curved at the theoretical values of ER, i.e. 100% and 85%. Accordingly, considering the relationship between W_i/W_s and N_s/N_i is linear and the line slope decreases with the decrease of the efficiency ratio ER is an acceptable practical approximation.

The six curves of Fig. 3 have two functions. First, they explain the problem behavior and the factors affecting it, descriptively. The second is the correction of the measured N_m values due to the use of non-standard hammer weight, numerically. If the standard hammer weight is W_s and the wrong used hammer weight is W_i. The obtained misleading blow counts of using W_i are N_i. Based on the actual energy ratio, the appropriate curve and the value of W_i/W_S the N_S could be calculated. This N_S is the corrected value of the well-known term N_m. This is a simple and direct procedure to solve major field problem.

2.4   Gap, Hypothesis and Research Question

The literature review does not show any previous study suggested a convinced procedure to correct the measure blow counts N_m, if the used hammer weight or its drop height is non-standard. Also, the suggested equations such as Eqs. 3 and 4 are not closed form nor empirical or based on physical models. They based on logic. However, there existed nothing better and this gap needs to be closed.

It is believed here-in, that the overall rig efficiency affects the measured blow counts N_m. As the efficiency decreases i.e. old machines, N_m increases, i.e. overestimated N_m, it is non-seen harm. This hypothesis needs to be verified. It could not be avoided, how it could be corrected?

Based on the above this research has been planned to answer the following questions:

1. If the hammer weight W_S or its drop height H_S is reduced to its half, will N_m be doubled? As an example!
2. Is the relationship between W and N_m linear? If so, is the line slope function of the rig efficiency?
3. Is it reliable to use a physical model as a strategy and method to answer the above inquiries?

3   Methodology Method and Tool

Two expressions, methodology and method are closely connected and serving two different tasks. The methodology is the strategy that outlines the way in which research to be carried out. Method is the used tool to create the data needed to answer research question. With one methodology, several methods may be used to compensate research hypothesis and achieve its goal.

3.1   Methodology or Thesis Strategy

The paper main idea is dealing with SPT machine uses non-standard parameters. Namely they are hammer weight W_f and/or hammer drop height H_f. This definitely produces incorrect measured blow counts N_f. Is it possible to suggest empirical equations or charts to correct back the N_f to its correct value N_{Std}? The logical strategy should be experimental and the method is either a full scale or physical model. Here-in, we go to physical model, and the question is why?

The physical model is an early method used to simulate the behavior of a full-scale object by building one to-scale. Its function is not to predict proportional values, due to the scale effect. But, its results are very representative, when it comes to prediction of empirical relations or behavior explanation. The questions are what is the trend of the curve related different hammer weights W_f or hammer drop height H_f with the corresponding blow count N_f? Moreover, dose the variation of machine efficiency induce a family of curves? This needs several SPT-machines with different efficiencies, and each having capability of changing the hammer weight and hammer drop height.

Furthermore, the soil media must be homogeneous and constant to eliminate its variation effect on the tests results. Research wise, it is believed to use physical model during this study and one day field verification could be done.

3.2   Physical Model: Description

Figure 4 shows 2D and 3D "Solid Work software" schematic of the model. The model description had been fully reported and explained by Khater et al. (2019a, b), Turkey and Canada. Briefly as a reminder, the model consists of ten main parts; namely they are frame, sand tank, driven tube, smooth roller bearings, frictional-breakers, anvil, guides, hammers, load cell, digital converter and others minor parts and tools.

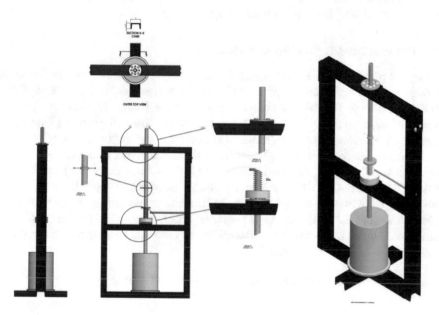

Fig. 4. The physical model 2D – 3D "Solid Works" schismatic

The model idea is; a hammer weigh W_i lifted up manually to a pre-defined constant height "H". Then, it freely falls down under its own weight repeatedly knocking a steel rode several knocks until pushing it inside the sand tank a certain distance "h". The number of knocks is the bow counts N_{mi}. Then, W_i changes to be W_{i+1} and so on and so far, the test repeated for five different W's. The same process has been repeated if the desired study is changing the hammer drop height H. The hammer weight W is kept constant, but the height of drop H_i changes five times. The number of blow counts N_{mi} has been reported per each value of H_i. The model efficiency controller has ability of changing efficiency level. A six different efficiencies have been considered, 100%, 85%, 70%, 60%, 50% and 35% they covers the maximum, minimum possible efficiency limits and the in between per each W_i and/or H_i.

The driven tube that stand for SPT-spoon, is made of steel has internal/external diameters of 15.71/22.86 mm, respectively with an arrow shape. Multi-function movement guides have been used. They are frictionless roller bearings that used during 100% efficiency. Later on, they replaced by manual frictional breakers to activate increasing resistance that gradually reduces the efficiency to the desired levels, in stages down to 35%. Two pair of these multi-function kits has been mounted on top and at middle of the model frame to secure the tube verticality. Also, guides have been soldered to the driven tube to get full control on the changes of hammer fallen height H_i and the tube penetration length. A load cell, having pair of strain gages, Wheatstone bridge and digital converter are used as quality control kit. It has been tested and calibrated. During every test once hammer hits the anvil, the data-logger records a reading to observe the model behavior. It counts number of blows, recording the rate of blows and observing the applied force variation, if any.

3.3 Physical Model: Efficiencies Adjustment

This article presents the used method to fine-tune the efficiencies. First, the frictionless bearings are attached, then a hammer weighs 9.12 N freely drops a height of 20.0 cm several times until the driven tube penetrates the sand tank by 15.0 cm. The total number of blow counts is N_{th}, i.e. at efficiency 100%. Then, the smooth bearings are replaced by frictional breakers. These breakers have the ability to apply adjustable lateral pressure with uniform roughness along the surface area of the driven tube. The total number of blows N_m has been increased to achieve the same penetration depth, i.e. 15.0 cm. The weight W, the drop height H and the penetration depth L are the same in both cases, i.e. with and without friction, but the consumed energy for penetrating the same depth is not the same because the number of blows increased for case of frictional bearing. To equate the two energies, the energy of frictional bearing case multiplied by a factor $\eta_K < 1.0$, see Eqs. 6 and 7 below:

$$W \times H \times N_{th} = \ell_K \times W \times H \times N_m \tag{6}$$

$$\ell_K = \frac{N_{th-total}}{N_{m-total}} \leq 1.0 \tag{7}$$

Here ℓ_K is the system global efficiency, numerically equivalent to ER. By simple try and error the applied lateral pressure is adjusted several times until $\ell_K = 0.85$, i.e. efficiency 85%. The same procedures repeated to all the used efficiencies, that is to say, 100%, 85%, 70%, 60% 50% and 35%.

Physical Model: Experiment Procedures

The sand tank is filled with compacted sand of thin layers to create a homogeneous soil mass. The sand reached the possible maximum dry density 18.40 kN/m³. This density kept constant during the rest of the study. The efficiency controller adjusted to the desired efficiency of each case study. Next, the chosen hammer that weighs W_S is mounted at a height of H_i cm a head of the driven tube upper tip. This height is kept constant during each individual case of study and re-adjusted back after every blow. However, the hammer keeps knocking the tube tip repeatedly until the tube penetrates the sand by 7.50 cm; at this stage, no number of blows is counted. Then, the guide is re-adjusted to let the tube penetration the sand tank a new depth equal to 15.0 cm. The test continues and the 15 cm penetration depth is achieved and the number of blows is counted and reported as, N_i. For accuracy, the penetration depth corresponding to every single knock has been measured too and the load cell results recorded to maintain the rate of blows constant. Also, every single experiment has been audio and video taped. The rate of the hammer dropping was seven blows per minutes, in-average. Then H_i is increased to be H_{i+1} and the test repeated five times to cover the cases of study for the same efficiency ratio η_K. Then the efficiency ratio η_K increased six times and the test repeated for the same range of H's, i.e. five drop heights. Accordingly, a total number of 30-cases have been carried out.

4 Data Generation and Cases of Study

The current study theme and focus are the effect of using non-standard hammer drop height on the measured blow counts N_m. Accordingly, the cases of study shown in Table 1, presents the data needed to be generated to perform the analysis later on. However, thirty cases of studies have been suggested to create qualitative data style. As shown in Table 1, five different hammer drop height ratios have been changed with each efficiency ratio. The base drop height is H_S and it is equal to 20.0 cm. It plays the role of standard drop height, while H_i presents the non-standard hammer drop heights. The efficiency ratio η_K or ER has been changed six times same as the previous performed studies, i.e. 100%, 85%, 70%, 60%, 50% and 35%. The used hammer weight $W_S = 9.12$ N and was kept constant during the study. Due to the axes normalization system, the numerical value of W_S does not affect the results, Fig. 5.

Table 1. Drop height changing ratios H_i/H_S vs. efficiency η_K

	Cases study					
	η_K					
H_i/H_S	100%	85%	70%	60%	50%	35%
	A	B	C	D	E	F
0.13	C11	C12	C13	C14	C15	C16
0.25	C21	C22	C23	C24	C25	C26
0.50	C31	C32	C33	C34	C35	C36
0.75	C41	C42	C43	C44	C45	C46
1.00	C51	C52	C53	C54	C55	C56

H_i/H_o = hammer drop height changing ratio;
Cii = case study number ii
η_K = efficiency ratio ER, % and A:E, curve titles, Fig. 4.

4.1   Soil Sample Properties

This section introduces the results of the performed physical and mechanical laboratory routine tests used to identify the soil that used to fills the sand tank of the physical model. As a start, the natural sandy soil has been used to create a reconstituted well graded ideal sandy soil mass, SW. Next, the physical and mechanical properties of the reconstituted samples have been determined, measured and then used during the course of this study. All the laboratory tests were per ASTM Book of Standards. However, the tests results are reported in Table 2, below.

Table 2. Reconstitutes soil sample properties, SW

[a]Sieve number	16	20	30	40	50	100	200
Opening, mm	1.18	0.85	0.6	0.43	0.30	0.15	0.075
% Retained by weight	40	18	12	8	6	8.5	7.5
[b]Sieve constants	$D_{10} = 0.19$, $D_{30} = 0.6$, $D_{60} = 1.18$, $C_U = 6.21$, $C_C = 1.81$, $G_S = 2.66$						
[c]Densities	$\gamma_{d\text{-(max)}} = 19.1$, $\gamma_{d\text{-(min)}} = 16.4$, $\gamma_{d\text{-(used)}} = 18.4$ $D_r = 0.77$						
Shear parameters	$\phi = 36.7°$ with $\gamma_{d\text{-(used)}}$, $\alpha = 29°$ with $\gamma_{d\text{-(min)}}$						

[a]ASTM system, [b]D's in mm, [c] γ's in kN/m^3.

Constructing a reconstituted sample, means re-accommodating pre-determine weight of specific particle diameter in its correct sieve. Table 2 shows the retained weight on each sieve to satisfy SW conditions. The authors suggest denoting this sand "Fayoum" sand, Fayoum is the city where the sample abstracted. This suggestion has two benefits, for researchers using numerical analysis the given values are trustable, for researchers using experimental techniques Table 2 could be used as is, and save a lot of time. Fayoum is a Pharaohs name and the Pyramids foundation soil is sand.

5   Presentation of Results

Table 1 shows, six efficiencies ER have been physically modeled. They are 100%, 85%, 70%, 60%, 50% and 35%. Figure 5 shows the results of the same efficiency ER are connected to build-up an experimental curve consists of five measured points presented as the best fit, Table 1. These points indicate five different hammer drop heights of a fallen hammer presented as ratios. However, x-axis co-ordinates are H_i/H_s, y-axis co-ordinates are N_s/N_i where i increases from 1 to 5. H_s is the major used height while N_s is the lowest blow counts, which all the time corresponding to H_s. This normalization system generalizes the use of the curves and also makes the ratios ranges from 0.0 to 1.0 this is practical and logical range.

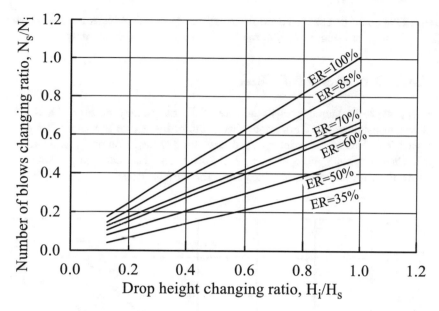

Fig 5 Blows counts changing ratio vs. hammer drop height changing ratio

5.1 Reading the Curves

Figure 5 displays six curves present efficiency ratios ER of 100%, 85%, 70%, 60%, 50% and 35% in descending order. In all cases, as the ratio H_i/H_s decreases, N_s/N_i decreases too. This says when the used hammer drop height H_i is shorter than the standard one H_S the measured N_i is over estimated and misleading, hammers drop height wise. Same observation spreads for efficiencies, as ER decreases; N_i increases too, efficiency wise.

Generally, for the same H_i/H_s if ER decreases N_s/N_i decreases too, i.e. N_i increases. Also, at H_i/H_s equal to 1.0, i.e. standard drop height used, if ER decreases N_s/N_i decreases, N_i increases and overestimated. This is over looked and non-seen harm.

The relationship between H_i/H_s and N_s/N_i is linear for all values of ER even for the value of ER = 100%. The slope of each curve is function of its efficiency. Accordingly, the relationship between H_i/H_s and N_s/N_i is linear and the line slope decreases with the decrease of the efficiency ER is a fact.

5.2 Advantages of Fig. 5

The six curves drafted within Fig. 5 have two advantages. First, they explain the behavior of using non-standard drop height problem and the factors affecting it, descriptively. The second and main advantage is the possibility to correct the measured N_m after the use of non-standard hammer drop height, numerically.

Assume the standard hammer drop height is H_S and the incorrect one is H_i. The obtained value of blow counts induced due to using H_i is N_i. Based on the energy ratio ER pick corresponding curve, calculate H_i/H_S and find N_s/N_i, y-axis. As N_i is known, corrected N_S could be calculated from the ratio of N_s/N_i. This N_S is the one known as N_m. It is simple procedure to solve major problem, even approximately.

6 Analysis of over All Results

To recap, two main figurers present the core of this paper; they are Figs. 3 and 5. They are direct measurements. Throughout this subtitle, the above measured values have been analyzed. Hence, the comings up figures drafted based on analysis of measured values. They are four figures, Figs. 6, 7, 8 and 9. This connects the previously published results with the results obtained from the current paper.

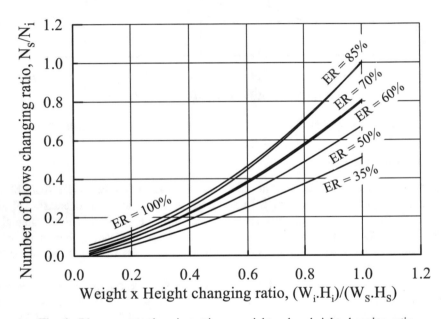

Fig. 6. Blows counts changing ratio vs. weight x drop height changing ratio

To draft Fig. 6, start with the smallest weight W_j then changing the height of drop H_i five times, i = 1:5 gives five values of N_{ji}. Then, multiply Wj by each value of H_i to get five values of $W_j.H_i$ having five corresponding values of Nji. This pair of five values ($W_j.H_i$; N_{ji}) are the x-axis and y-axis co-ordinates, of an energy ratio ER_k curve, where k = 1:6. Then, this process changed six times and Fig. 6 drafted.

The trend of Fig. 6 is a copy of Fig. 3 trend and close to Fig. 5, but Fig. 5 is more linear. Also, The trend and the values of ER = 100% and ER = 85% are nearly the same. Moreover, the trend and the values of ER = 70% and ER = 60% are nearly the same. At low values of ER, the curves tend to be linear, ER = 35% is pure line. The following six equations presents the imperial equations of the different energy rations presented in Fig. 6 based on the best fit technique, where R is the assurance factor:

Case of ER = 100%

$$\frac{N_S}{N_i} = 0.6072\left(\frac{W_iH_i}{W_SH_S}\right)^2 + 0.3606\frac{W_iH_i}{W_SH_S} + 0.031; R = 0.99 \tag{8}$$

Case of ER = 85%

$$\frac{N_S}{N_i} = 0.6787\left(\frac{W_iH_i}{W_SH_S}\right)^2 + 0.3023\frac{W_iH_i}{W_SH_S} + 0.022; R = 0.99 \tag{9}$$

Case of ER = 70%

$$\frac{N_S}{N_i} = 0.4322\left(\frac{W_iH_i}{W_SH_S}\right)^2 + 0.3630\frac{W_iH_i}{W_SH_S} + 0.005; R = 0.99 \tag{10}$$

Case of ER = 60%

$$\frac{N_S}{N_i} = 0.3903\left(\frac{W_iH_i}{W_SH_S}\right)^2 + 0.4256\frac{W_iH_i}{W_SH_S} + 0.009; R = 0.99 \tag{11}$$

Case of ER = 50%

$$\frac{N_S}{N_i} = 0.3031\left(\frac{W_iH_i}{W_SH_S}\right)^2 + 0.3845\frac{W_iH_i}{W_SH_S} + 0.0148; R = 0.99 \tag{12}$$

Case of ER = 35%

$$\frac{N_S}{N_i} = 0.1933\left(\frac{W_iH_i}{W_SH_S}\right)^2 + 0.3382\frac{W_iH_i}{W_SH_S} + 0.021; R = 0.98 \tag{13}$$

6.1   Normalization with Respect to Efficiency Ratio

As show above, Fig. 6 has six curves and six empirical equations. Somehow, this is complicated if we deal with Figs. 3, 5 and 6 altogether. Accordingly, the following three figures, i.e. Figs. 7, 8 and 9, are simplifications of Figs. 3, 5 and 6 with respect to the energy ratio ER, i.e. "≈" normalization. Table 3 presents one example explains the method used for calculations within Fig. 7, the rest are the same, i.e. Figs. 8 and 9 calculations. It is needless to say the, simplification reduces the assurance factor, but to acceptable range.

$$\frac{N_S}{N_i} = 0.98\left(\approx \frac{H_i}{H_S}\right) + 0.102 \tag{14}$$

$$\frac{N_S}{N_i} = 1.12 \left(\approx \frac{W_i}{W_S} \right) - 0.04 \qquad (15)$$

$$\frac{N_S}{N_i} = 1.10 \left(\approx \frac{W_i H_i}{W_S H_S} \right) - 0.04 \qquad (16)$$

Table 3. Example of Fig. 5 normalization to produce Fig. 7

Case of H_i = 2.5 cm, as an example Has been repeated for H_i = 5.0, 10.0, 15.0 and 20.0 cm						Y axis	X axis
H_i	H_s	η	H_i/H_s	N_i	N_s	$\eta^* H_i/H_S$	N_S/N_i
2.5	20	1.00	0.125	93	14	0.151	0.151
2.5		0.85	0.125	107		0.111	0.131
2.5		0.70	0.125	147		0.067	0.095
2.5		0.60	0.125	159		0.053	0.088
2.5		0.50	0.125	194		0.036	0.072
2.5		0.35	0.125	365		0.013	0.038

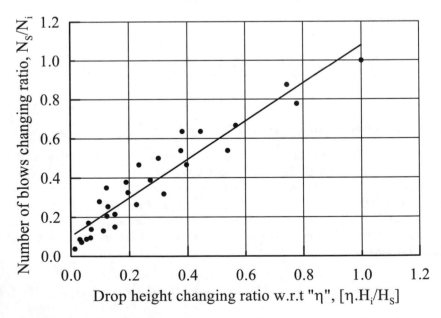

Fig. 7. Blow counts changing ratio vs. "$\approx$" x drop height changing ratio

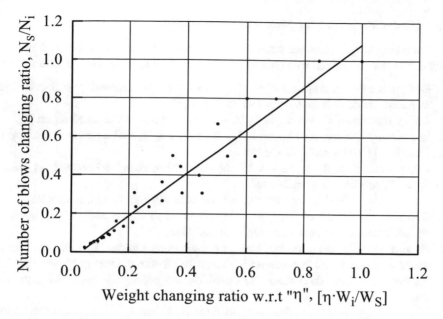

Fig 8 Blows counts changing ratio vs. weight "$\mathscr{C}$" x weight changing ratio

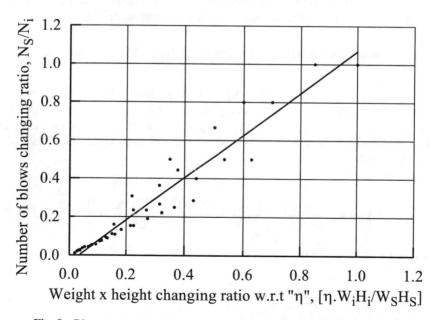

Fig. 9. Blow counts changing ratio vs. "$\mathscr{C}$" x weight-height changing ratio

6.2 Khater Procedure for Measuring Efficiency

The following simple steps are strongly recommended by authors to be used in every project and it is suggested to be added to SPT standard specifications within codes:

1. Perform the SPT test as normal and get N_S which is corresponding to H_S, one time as a start inside only one bore hole at any depth.
2. Adjust the hammer drop height to $H_i < H_S$ and repeat the test and report N_i.
3. Repeat step 2 twice, for H_{i+1} and H_{i+2} both H_i's $< H_S$ and report N_{i+1} and N_{i+2}
4. Calculate N_i the average of the three.
5. Apply Khater's equation, i.e. Eq. 16, the only unknown is "℮" because Ns, Ni and $W_i = W_s$ are known or calculated.
6. The calculated "℮" is the SPT overall machine efficiency, approximately.
7. Based on the calculated "℮" go to Eqs. 8 to 13 of Fig. 6 and pick up the most close efficiency and hence its empirical equation.
8. Perform the SPT test as normal based on the regular standard specifications for all the project bore holes then correct the measured in-site number of blows, here it is N_i, back to its standard value N_S based on the appropriate empirical equation, Eqs. 8 to 13.
9. This is an approximate method suggested by us; definitely it needs more refinement and development by further research work.
10. Kindly, any development of the above procedure or Eq. 16 should be denoted as modified Khater procedure or modified Khater equation, please.

6.3 Suggested Modifications for Burmister (1948) and Lacroix and Horn (1973)

It is suggested to modify Eq. 3, and Eq. 4 to be as shown below. However, these suggestions are important because the two given equations,i.e. Eq. 3, and Eq. 4 did not take the efficiency into consideration; it is very effective and must be considered. The above mention names are great scientists, we do not criticize, they are pioneers and we hardly try to follows their footsteps.

Modified Burmister (1948):

$$N_{cr} = 1.1 \, ℮ \, N_m \frac{(W \text{ lbs}) (H \text{ in})}{(140 \text{ lbs}) (30 \text{ in})} \frac{[(2.0 \text{ in})^2 - (1.375 \text{ in})^2]}{D_o^2 - D_i^2} - 0.04 \qquad (17)$$

Modified Lacroix and Horn (1973):

$$N_{cr} = 1.1 \, ℮ \, \frac{2 \, N_m \, W_i \, H_i}{175 \, D_i^2 \, L_i} - 0.04 \qquad (18)$$

6.4   Sorting-Out the Machine Efficiency

Based on the above study as well as Eq. 14, the drop height H_i could be increased to satisfy two conditions. First, $N_S/N_i = 1.0$, while the second is the desired "$\mathcal{e}$". Accordingly, the machine efficiency "$\mathcal{e}$" could be fixed, costless and in no time.

6.5   Important Remark

One may agree or disagree with the degree of accuracy of the obtained result of N_m. The results are obtained from physical model; this needs intensive filed tests for verification. Again, it is more or less an approximation, guide values and encouraging for further and future research work to be better improved.

7   Conclusions and Recommendations

This final heading contains three subtitles. The first is evaluation and self-criticism. The second is the conclusions and main findings. The third is recommendations and suggestion for practice and further research works.

7.1   Achievement or Complete Failure

At the end of the literature review, the authors stated the gap in literature that needs to be filled, the hypothesis and three research questions. It is needless to say, if one of them has not been answered and satisfied, it will be a complete failure.

The gap in literature shows, there is no convinced procedures exists to correct the measure blow counts N_m, where the used SPT parameters are non-standard. Here-in, the answer is, refer to Khater procedures and Eq. 16.

The hypothesis believes that the overall rig efficiency affects the measured blow counts N_m. Yes, as the efficiency decreases N_m increases, i.e. overestimated. True hypothesis and has been proved, Figs. 3, 5 and 6.

The first research question was; If the hammer weight W or drop height H has been reduced to its half, will N_m be doubled? The answer is no, not exactly, referring to Figs. 3 and 5. Second research question was; Is the relationship between W or H and N_m linear? The answer is yes for "H" and up-to for "W", slope of the line affected by machine efficiency. Third question was; Is it reliable to use a physical model as method to answer the above? Yes, perfect method and methodology.

The paper title is; equations to correct SPT-N values obtained using non-standard hammer weight and/or drop height. The text matches the title and the equations have been given also a novel procedure has been suggested. Moreover, modifications to previously developed two equations by others have been proposed, here-in.

7.2   Conclusions and Main Findings

1. A novel procedure has been suggested "Khater procedures" to measure the SPT-Rig efficiency at the field, on spot.

2. Equations have been suggested to correct the measured N_m.
3. Suggesion given to adjust machine efficiency by increasing H_i.
4. The rig overall efficiency ER greatly affect the measured number N_m.
5. As the rig efficiency decreases, measured number of bows N_m increases thus misleading, i.e. over estimating the soil strength and deformation properties.
6. The relationship between the hammer weights W and/or the drop height H and the measured number of blow counts N_m is linear, and the slope of the line decrease as the efficiency decreases. This leads to over estimating N_m.
7. Incorrect number of blows that induced from the uses of non-standard SPT parameters could be corrected back to its standard one sees Khater procedure.
8. It is strongly recommended to force by the power of low every SPT-Machine to have regular calibration and must be certified and the ER must be reported in every geotechnical reports used this rig.
9. It is strongly recommended to apply the suggested method for machine efficiency prediction per every site and must be compared to the certified calibrated value, preferable to add this conclusion to the standard specifications codes, at least as a guide value.
10. The physical model is effective to explain the behavior of such problem. Its numerical results could be used if they are presented in dimensionless forms.

7.3 Further Research Work

It is suggested to extend this study to be intensive field work to consolidate and verify the given conclusions, suggested procedures and equations. Also, if the finite element analysis is used as a 3D-dynamic problem to investigate the same cases studied here-in. This could be a new contribution.

Acknowledgements. The authors present their deep gratitude to Prof Dr. A. Bazaraa for his kind personal communication, direct input and advices during discussions on topic.

References

Aggour, M.S., Radding, W.R.: Standard Penetration Test (SPT) Correction. Technical Report. Civil and Environmental Engineering Department, University of Maryland (2001)

Ameratunga, J., Sivakugag, N., Das, B.M.: Correlations of Soil and Rock Properties in Geotechnical Engineering. Springer, India (2016)

ASTM, 1999, D 1586-99. Standard Test Method for Penetration Test and Split-Barrel Sampling of Soils. American Society for Testing and Materials. Annual Book of ASTM Standards, Philadelphia (1999)

Bazaraa, A.R.S.: Use of standard penetration test for estimating settlement of shallow foundations on sand. thesis presented to University of Illinois, in partial fulfillment of the requirements for the degree of Doctor of Philosophy (1967)

British Standards: Determination of the penetration resistance using the split barrel sampler. The standard penetration test, SPT. B.S.1377 Part 9 (1990)

Burmister, D.M.: The importance and practical use of relative density in soil mechanics. Proc. ASTM **48**, 1249–1268 (1948)

Butler, J.J., Caliendo, J.A., Goble, G.G.: Comparison of SPT energy measurement methods. In: Proceedings of International Conference on Site Characterization, ISC 1998, Balkema, Rotterdam, The Netherlands, pp. 901–906 (1998)

De Mello, V.F.B.: The standard penetration test. In: In Proceedings of the 4th Pan-American Conference on Soil Mechanics and Foundation Engineering, San Juan, PR, vol. 1, pp. 1–86 (1971)

Fang, H.Y.: Foundation Engineering Handbook. Kluwer Academic Publishers, Boston (1999)

Farrar, J.A.: Summary of standard penetration test (SPT) energy measurement experiment. In: Proceedings of International Conference on Site Characterization, ISC 1998, Balkema, Rotterdam, The Netherlands, pp. 919–926 (1998)

Fletcher, G.F.A.: Standard penetration test: its uses and abuses. J. Soil Mech. Found. Div. **91**(4), 67–75 (1965)

Howie, J., Campanella, R.G.: Energy measurement in the standard penetration test (SPT). Department of Civil Engineering, University of British Columbia, Vancouver, BC, Canada (2008)

Khater, R.Kh.: Correction factor to enhance the non-standard SPT hammer effect. In: 18th International Conference on Soil Mechanics and Geotechnical Engineering, World Academy of Science (2016)

Khater, R.Kh.: The SPT- theory practice and correlations. Text book style, under publication stage (2018)

Khater, R.Kh., Abdelrahman, G.E., Baset, M.A.: Equations to Correction SPT N-Values Obtained from Using Non-Standard Hammer Weight – Part I, International Civil Engineering and Architecture Conference, Karadeniz Technical University, Trabzon, Turkey (2019a)

Khater, R.Kh., Abdelrahman, G.E., Baset, M.A.: Equations to Correction SPT N-Values Obtained from Using Non-Standard Hammer Weight – Part II. Paper ID: 189. GeoSt.John's 2019. In: The 72nd Canadian Geot. Conference, Newfoundland and Labrador, Canada (2019b)

Lamb, R.: SPT Hammer Calibration Update. A Memo to Geotechnical Engineering Section, Minnesota Department of Transportation (2000)

Look, B.G., Seldel, J.P., Sivakumar, S.T., Welikala, D.I.C.: Standard penetration test measurement variations exposed using a digital PDM device. In: International Conference on Geotechnical Engineering, ICGE, Colombo, pp. 451–454 (2105)

Peck, R.B., Bazaraa, A.R.S.: Discussion on settlement of spread footing on sand. J. Soil Mech. Found. Div. **95**(5), 905–909 (1969)

Skempton, A.W.: Standard Penetration Test Procedures and the Effect in Sands of Overburden Pressure, Relative Density, Particle Size, Aging and Over consolidation. Geotechnique **36**(3), 425–447 (1986)

Sherbiny, R.M., Salem, M.A.: Evaluation of SPT Energy for Donut and Safety Hammers Using CPT Measurements in Egypt. Cairo University, Egypt (2013)

Tokimatsu, K., Seed, H.B.: Evaluation of settlements in sand due to earthquake shaking. J. Geotech. Eng. **113**(8) (1987)

Tsai, J.S., Liou, Y.J., Liu, F.C., Chen, C.H.: Effect of hammer shape on energy transfer measurement in the standard penetration test. Soil Found. **44**(3), 103–114 (2004)

Youd, L., et al.: SPT hammer energy ratio versus drop height. J. Geotech. Geoenviron. Eng. (2008)

Experimental Evaluation of Friction Losses Between Post-tensioned Strands and Plastic Ducts

Nour El Deen El Ezz[1(✉)], Hisham Basha[2], and Zaher Abou Saleh[3]

[1] Civil and Environmental Engineering, Rafik Hariri University,
Meshref, Lebanon
elezznm@students.rhu.edu.lb
[2] Engineering, Rafik Hariri University, Meshref, Lebanon
bashahs@rhu.edu.lb
[3] Rafik Hariri University, Meshref, Lebanon
abousalehza@rhu.edu.lb

Abstract. Friction losses are short term losses in Post-Tensioned systems. The friction created at the interface of tendons and ducts during the stretching of a curved or straight tendon in a post-tensioned member leads to a drop in the prestress along the member from the stretching end. The purpose of this research is to study the friction losses between tendons and plastic ducts. Moreover, it has been shown that "corrugated steel ducts will fracture at cyclically opening cracks across the duct. Plastic ducts have been shown to perform better under similar conditions" [2]. This proves that plastic ducts are good ducts in durability and corrosion protection. Thus, the friction parameters for plastic ducts system will be evaluated in this research paper. The analysis results presented that the average wobble coefficient and the average curvature one for plastic ducts of 0.005/m and 0.1/rad, respectively, which are less than the values for steel ducts. Plastic ducts reduce the friction losses to 41% in comparison with the steel ducts at site work. In addition, plastic ducts reduce the friction losses to 33% in comparison with the steel ducts at laboratory work as presented in this paper. The values of (k) and (μ) for only metal sheathings, rigid metal ducts, and other types are introduced in PCI (Fig. 2)³. The primary goal of this research was to conduct an experimental evaluation of the friction losses between post-tensioned strands and plastic ducts. Moreover, in the (Fig. 14) below, the values of (k) and (μ) for only metal sheathings, rigid metal ducts, and other types are introduced [3]. In this research, the values for (k) and (μ) will be evaluated for plastic ducts in order to introduce values for (k) and (μ) for plastic ducts in the (Fig. 2).

1 Introduction

In prestressed concrete field, one of the most important factors is to determine the effective prestressing force after all losses. In the past, researchers noticed that the value of prestressing force didn't stay constant [5]. At transfer and service stages, there is a drop-in prestress forces along with the strands in post tensioned system. This study

© Springer Nature Switzerland AG 2020
H. Shehata et al. (Eds.): GeoMEast 2019, SUCI, pp. 166–183, 2020.
https://doi.org/10.1007/978-3-030-34184-8_11

focuses on measure of the friction losses between strands and plastic ducts in post-tensioned system. It is important to determine the friction coefficients such as the curvature coefficient (μ) and the wobble coefficient (k) for plastic ducts through laboratory and site work (Fig. 1).

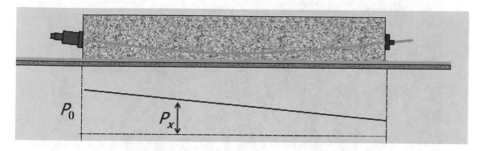

Fig. 1. Variation of prestressing force after tensioning [5]

1.1   Literature Review

The values of (k) and (μ) for only metal sheathings, rigid metal ducts, and other types are introduced in PCI (Fig. 2)[3]. In this research, the values for (k) and (μ) will be evaluated for plastic ducts. Corrugated plastic ducts are more durable ducts and they are protected against corrosion (Fig. 3).

Table 3.7 Wobble and Curvature Friction Coefficients

Type of tendon	Wobble coefficient, K per foot	Curvature coefficient, μ
Tendons in flexible metal sheathing		
Wire tendons	0.0010–0.0015	0.15–0.25
7-wire strand	0.0005–0.0020	0.15–0.25
High-strength bars	0.0001–0.0006	0.08–0.30
Tendons in rigid metal duct		
7-wire strand	0.0002	0.15–0.25
Mastic-coated tendons		
Wire tendons and 7-wire strand	0.0010–0.0020	0.05–0.15
Pregreased tendons		
Wire tendons and 7-wire strand	0.0003–0.0020	0.05–0.15

Source: Prestressed Concrete Institute.

Fig. 2. Values of coefficient of friction [3]

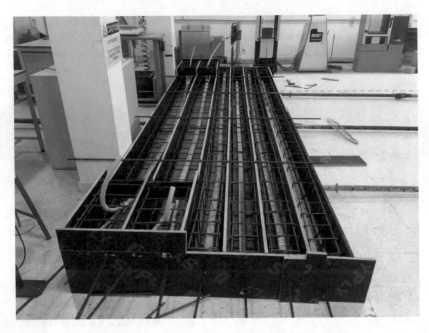

Fig. 3. The presence of plastic ducts and steel ducts

This research paper divides the work into steps. The first step is to estimate the wobble coefficient between the tendon and duct in one straight tendon. In first step, two samples will be used. The first sample consists of one straight tendon in plastic duct. The first sample will help in determining the (k) of the plastic duct. While, the second sample consists of one straight strand with steel sheathing duct. It is essential to determine the difference between the wobble coefficients of steel and plastic ducts in this research. The existing of one tendon makes elastic shortening in strand equals to zero. In addition, the straight tendon makes (α) equals to zero. Hence, the first step evaluates the wobble coefficient (k). The load cells will measure the force in the tendon after jacking the strand in the first step, wobble coefficient (k) will be evaluated by this equation:

$$[\Delta f_p]_F = f_{pj}[\mu\alpha + kx] \tag{1}$$

$$\int_{p0}^{px} \frac{dp}{p} = -[\mu \int_0^\alpha d\alpha + k \int_0^x dx] \tag{2}$$

The solution of the above differential equation is

$$\ln \frac{P_x}{P_0} = -(\mu\alpha + kx)$$

Then,

$$P_x = P_0 e^{-(\mu\alpha + kx)} \tag{3}$$

$$Px = P0\left[1 - (\mu\alpha + kx)\right] \tag{4}$$

(P0) is the prestress at the stretching end after any loss due to elastic shortening. For small values of $(\mu\alpha + kx)$, the derived Eq. (3) can be simplified by the Taylor series expansion to be $\{Px = P0\,(1 - (\mu\alpha + kx)\}$. Taylor expression for $ex = £$ $xn/n! = 1 + x + x2/2! + x3/3!......$ So, if x is replaced by $(\mu\alpha + kx)$, then this Eq. (4) will be proved $e - (\mu\alpha + kx) = 1 - (\mu\alpha + kx)$. By using the first degree of the Taylor equation, the Eq. (4) was well proved. From the simplified Taylor equation, the variation of the prestressing force is linear with the distance from the stretching end. For example, P1 = P0e − ($\mu\alpha$ + kx), this means as proved before, P1 = P0(1−($\mu\alpha + kx$)). Where (P1) and (P0) are the prestressing forces at two different points in the strand. By using Eq. (4), the force can be determined at any position in the tendon.

$[\Delta f_p]_F$ will be calculated in kN and it is the difference between live load cell and dead load cell measurements after stretching of the tendon in the duct (Fig. 4).

Fig. 4. Dead and live end anchors with load cells are connected through data acquisition

The jacking force at live end was calibrated by the load cell measurement at live end. In addition, all strands are jacked at 14 tons from the live end.

The second step is to evaluate the curvature coefficient for plastic and steel ducts. In the second step, two other samples are used. Sample (3) consists of one draped strand with steel sheathing duct while sample (4) consists of one draped strand with plastic duct. This step evaluates the curvature coefficient (μ) for the two duct types. The (μ) coefficients for the two duct types will be determined from Eq. (1) and by knowing the (k) coefficients from the first step.

Table 1. Test samples for lab work

Number of samples	f'_c expected at transfer	Duct types	Number of strands	Tendon profile	Modulus of elasticity to be determined by experiment (MPA)	Concrete section properties (mm * mm)	Jacking force
Sample 1	25	Steel	1 strand	Straight strand	195000	200 * 300	14 tons
Sample 2	25	Steel	1 strand	Draped strand	195000	200 * 300	14 tons
Sample 3	25	Plastic	1 strand	Straight strand	195000	200 * 300	14 tons
Sample 4	25	Plastic	1 strand	Draped strand	195000	200 * 300	14 tons
Sample 6	25	Steel	2 strands	Draped strands	195000	300 * 400	14 tons
Sample 7	25	Plastic	2 strands	Draped strands	195000	300 * 400	14 tons

The variation of samples in laboratory work will be summarized in (Table 1).

1.2   Research Significance

The primary goal of this research was to conduct an experimental evaluation of the friction losses between post-tensioned strands and plastic ducts. Moreover, in the (Fig. 14) below, the values of (k) and (μ) for only metal sheathings, rigid metal ducts, and other types are introduced [3]. In this research, the values for (k) and (μ) will be evaluated for plastic ducts in order to introduce values for (k) and (μ) for plastic ducts in the (Fig. 2).

1.3   Experimental Investigation

One load cell was set on live end and another one was set on the dead end. The difference between live end and dead-end load cells is the drop-in force resulted from friction losses. In lab, the samples are of length 5 m. Post tensioned beams started from this length. While, in site, the length for beams is about 10 m. To take the effect of length on the friction losses. The results indicate that plastic ducts reduce the friction losses in comparison with steel ducts (Tables 2 and 5) (Fig. 5).

Table 2. Summary for the friction losses values for all types of beams in lab

Type of beams for 4.8 m samples	$\left[\Delta f_p\right]_F$ in KN for friction losses
PTB-S-P	3.2
PTB-S-S	9.81
PTB-C-P	10.7
PTB-C-S	16.09

Table 3. (k) and (μ) values for the different types of beams used in the lab

Type of beams	k/m	μ/rad
PTB-S-P	0.00445	0
PTB-S-S	0.014057	0
PTB-C-P	0.00445	0.1739
PTB-C-S	0.014057	0.2
PTB-C-2S	0.014057	$0.1<\mu<0.2$
PTB-C-2P	0.00445	$0.29<\mu<0.32$

Table 4. Friction losses values at site

Type of beams	k/m	μ/rad
PTB-S-S	0.00872	0
PTB-S-P	0.00625	0
PTB-C-P	0.00625	0.012
PTB-C-S	0.00872	0.158

Table 5. Coefficients of (k) and (μ) for different types of beams at site

Type of beams for 9.2 m samples	$\left[\Delta f_p\right]_F$ in KN for friction losses
PTB-S-P	7.295
PTB-S-S	11.50757
PTB-C-P	12.618
PTB-C-S	21.754

1 Materials. The concrete used for casting beams in laboratory and in site had a design strength of 5083 psi (35 MPa) and was supplied by a local ready-mix plant (Tables 3 and 4).

Fig. 5. Measurement of post-tensioned beam with straight plastic duct using load cells

The prestressing strands used were 0.5 in. (12.7 mm) diameter seven wire strands conforming to ASTM standard A416, with a specified ultimate strength of 270 ksi (1861 MPa) and an average modules of elasticity of 28600 ksi (197128 MPa).

2 Fabrication of Specimens. In lab, the samples dimensions are described in the (Table 1). In site, the beams are put in slab of thickness 22 cm. The variety for plastic ducts and steel one was used in lab and in site in order to have full comparison between them.

1.4   Experimental Results

The results are provided in Tables 2 and 5. Friction losses in plastic draped ducts are decreased to 33% in comparison with the friction losses in steel ducts in Laboratory work for 5 m samples. In site, Friction losses in plastic draped ducts are decreased to 41% in comparison with the friction losses in steel ducts in Laboratory work for 5 m samples. Moreover, all the (k) and (μ) values for plastic ducts are less than the (k) and (μ) values for steel ducts.

2   Future Research

The results in this paper were based on the calculation of friction losses from the difference between live end and dead end. At the middle, it is recommended to put strain gauges sensors in order to assure that the friction loss relationship is

Fig. 6. Internal force measurement in strand using load cells during jacking of PTB-S_P

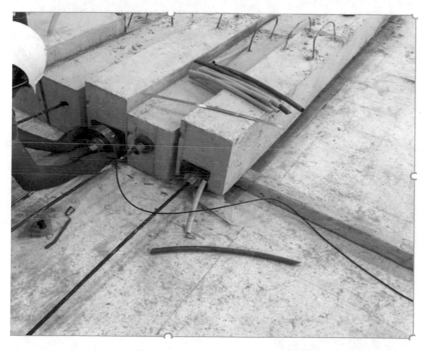

Fig. 7. The load cells are connected to the strand in straight steel duct

Fig. 8. Internal force measurement in strand using load cells during jacking of PTB-S_S

Fig. 9. Measurement of post-tensioned beam with curvature steel duct using load cell

Fig. 10. Internal force measurement in strand using load cells during jacking of PTB-C_S During Jacking

Fig. 11. Measurement of post-tensioned beam with curvature plastic duct using load cell

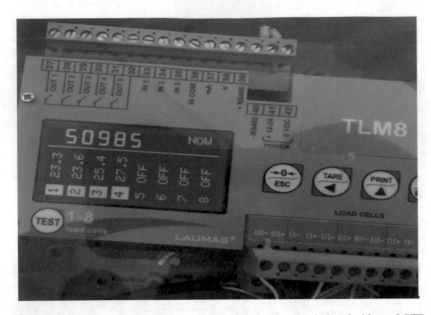

Fig. 12. Internal force measurement in strand using load cells during jacking of PTB-C_P during jacking

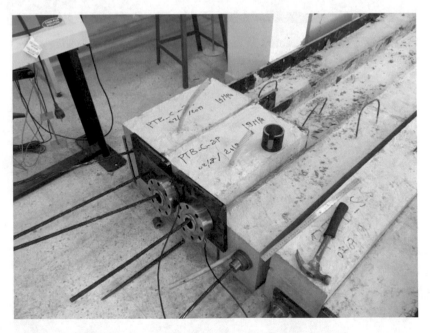

Fig. 13. Two load cells from each side are connected

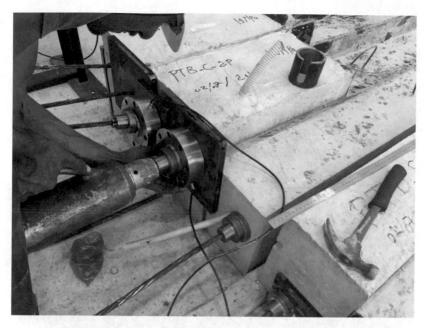

Fig. 14. Each strand was jacked alone starting from this strand

Fig. 15. Internal force measurement in strand using load cells of PTB-C-2P during jacking of the first strand

Fig. 16. Internal force measurement in strand using load cells of PTB- C-2P during jacking of the second strand

Fig. 17. Load cells are connected from two sides for the curved steel ducts

Fig. 18. Internal force measurement in strand using load cells of PTB-C-2S during jacking of the first strand

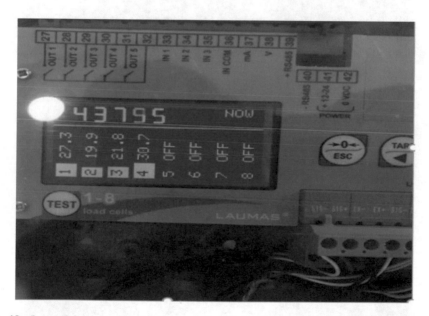

Fig. 19. Internal force measurement in strand using load cells of PTB-C-2S during jacking of the second strand

approximately linear. Beams of larger lengths may be introduced in order to take more

Fig. 20. Site work

Fig. 21. Load cells installation at site

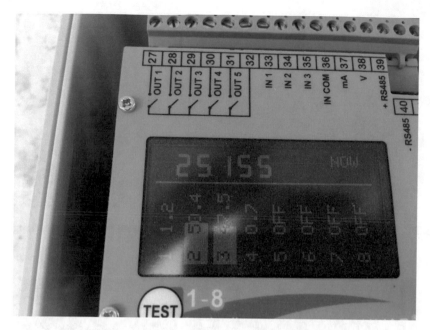

Fig. 22. Internal force measurement in strand of PTB-S-S using load cells during jacking at site

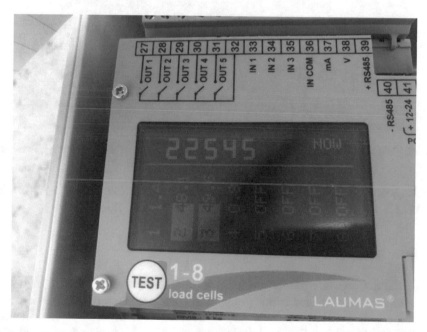

Fig. 23. Internal force measurement in strand of PTB-S-P using load cells during jacking at site

Fig. 24. Internal force measurement in strand of PTB-C-S using load cells during jacking at site

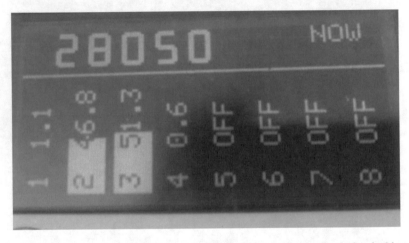

Fig. 25. Internal force measurement in strand of PTB-C-P using load cells during jacking at site

results about the effect of the length of the beam on the friction losses (Figs. 6, 7, 8, 9, 10, 11, 12, 13, 15, 16, 17, 18, 19, 20, 21, 22, 23, 24, and 25).

3   Conclusion

Plastic ducts reduce the friction losses in the Post-tensioned systems in comparison with steel ducts as presented in this paper. Plastic ducts are corrugated ducts in order to assure the bond between the ducts and the concrete. The (k) and (μ) coefficients for plastic ducts are less than the coefficients for steel ducts. This is an indication on having less friction losses by using plastic ducts. Plastic ducts reduce the friction losses to 41% in comparison with the steel ducts at site work. In addition, plastic ducts reduce the friction losses to 33% in comparison with the steel ducts at laboratory work as presented in this paper.

Acknowledgments. I would like to thank Dr. Zaher Abou Saleh for his full assistance and support in doing this research paper. Dr. Abou Saleh followed up with me all the steps until the finishing of this research paper.

Finally, I would like also to thank Rafik Hariri University which provide to us all the necessary funding to complete this research.

References

1. ACI Committee 318: Building code requirements for structural concrete (ACI 318-14), American Concrete Institute (ACI), Farmington Hills, MI, US (2014)
2. Granz, H.R., et al.: Corrugated plastic ducts for internal bonded post- tensioning. Task Group 9.6 Plastic Ducts of Fib Commission 9 (2000)
3. Nawy, E.G.: Prestressed Concrete: A Fundamental Approach (Fifth edition update ed.). The State University of New Jersey (2009)
4. Se-Jin Jeon, S.Y.-H.: Estimation of friction coefficients using smart strand. Int. J. Concr. Struct. Mater. 9(3), 369–379 (2015)
5. Sengupta, A.K., Menon, D.: Losses In Prestress (Part II). Indian Institute of Technology Madras, India

Behavior of Anchored Sheet Pile Wall

Pratish Kannaujiya and Vinay Bhushan Chauhan[✉]

Civil Engineering Department, Madan Mohan Malaviya University
of Technology, Gorakhpur 273010, India
kannaujiyapratish@gmail.com,
chauhan.vinaybhushan@gmail.com

Abstract. Present study discusses the behavior of an anchored sheet pile wall supporting a wide excavation based on the analysis using a computational tool based on the finite element method. The soil is modeled using the Mohr-Coulomb material and the sheet pile wall is modeled using the plate elements. A parametric study using numerical modelling is performed to investigate the behavior of an anchored sheet pile wall. The anchoring system is modeled by a combination of connectors (which do not interact with the soil) and geo-grids (which do interact with the soil and are used to account for grouting). The stability of the structure assessed using Strength Reduction Method and factor of safety for various combinations are calculated in the upper and lower bounds analysis. Results obtained from the parametric study indicate that the embedded depth of sheet pile wall, positioning of anchor, and length of anchors, inclination of the anchor govern the overall stability of structures studies in the present study. Moreover, the present study captures the failure modes of the anchored sheet pile wall system with the inclusion of multiple complex interfaces and supports conditions.

Keywords: Numerical modeling · Sheet pile · Anchored sheet pile · Finite element method

1 Introduction

Anchored sheet pile walls are used as either permanent or temporary structures used to retain the soil or water for a specific period of time, to build a structures like waterfront structures, cofferdams, cut-off walls under dams, erosion protection, excavation support system, and floodwalls etc. on the other side of the sheet pile wall (Bilgin and Erten 2009). The sheet pile structure consists of a continuously interlocked pile segments embedded in soils to resist horizontal pressures exerted by the backfill material. The sheet pile walls are constructed as a cantilever or anchored one depending on the height of excavation to be supported. Although cantilever sheet pile walls are usually provisioned to support the excavation height up to 12 m (Bilgin 1994; Dawkins 2001; and Bilgin 2010), but for higher wall requirement, anchors are provisioned to the sheet pile wall. To support the higher height of excavation, it is very common to use multiple anchors per unit length of the wall along with the height of the sheet pile wall. In this technique, anchors are installed along the height of the sheet pile wall to counteract against the lateral earth pressure exerted from the retained backfill. These anchors are

H. Shehata et al. (Eds.): GeoMEast 2019, SUCI, pp. 184–195, 2020.
https://doi.org/10.1007/978-3-030-34184-8_12

installed in such a manner that a smaller portion of the anchor length is bonded at one end in the soil, usually grouted or sort of a geogrid having a small length which holds the anchor in its position and another end of anchor provides anchoring force that works against earth pressure to transfer the pressure to the anchor head. This anchorage system leads to a retaining system where a thinner section of the sheet pile wall is required compared to a conventional cantilever sheet pile wall without anchors leading to an efficient and economic structure. Moreover, it requires lesser time to complete the excavation job with great efficiency. Furthermore, there are many cases where a very stiff ground (availability of rock strata or presence of gravel and boulders) may encounter at a shallow depth. Even due to the presence of existing underground construction or transmission lines, embedded depth of sheet pile needs to be curtailed at a shallow depth. In such cases, the anchored sheet pile wall provides a sustainable solution.

From the previous studies, it has been noted that provision of the anchorage system to sheet pile wall, enhances the overall stability of the structural system. Moreover, the stability is being controlled by many important parameters such as anchor's length, the position of the anchorage from the original ground surface, number of anchors per unit length of wall along its height, inclination of anchor with horizontal ground (θ) etc. These factors not only governs the overall factor of safety (FOS) of the system but also optimum selection of aforementioned parameters leads to an economic design of the anchored sheet pile wall system. So, a comprehensive numerical study based on finite element method has been carried to investigate the effect of anchor's length, the position of the anchor from the original ground surface, the number of anchors per unit length of the wall along its height and the inclination of the anchor with the horizontal ground on the sheet pile wall system.

2   Numerical Modeling of Anchored Sheet Pile Wall

In the present study, a finite element analysis was performed to evaluate the factor of safety (FOS) of the sheet pile wall system using OptumG2 (2017) based on the strength reduction method using upper and lower bound elements. The current study has been carried out to analyse the support system to an excavation of 10 m height (H) using a sheet pile wall considering the two-dimensional analysis. The analysis is carried out in two-dimensional analysis as the plane strain condition prevails in the problems like a sheet pile wall system. Boundary conditions are applied at the edges of the mesh i.e. fixed boundary condition at the bottom of the mesh and the roller boundary condition at the vertical edges of the soil mass. A typical mesh considered for this analysis with complete details of the geometrical configuration of a sheet pile wall is shown in Fig. 1.

Estimation of the optimum number of elements in the mesh is an essential part to execute the efficient analysis in terms of time-saving as well as better evaluation of outcomes that may be obtained from the analysis. In order to get the optimum number of total elements in the mesh considered for the study for the efficient execution of the analysis, a sensitivity analysis is carried out for the mesh shown in Fig. 1 by varying the total number of elements ranging from 2000 to 12000. Based on the above simulations, the factor of safety is determined for each simulation using upper bound and

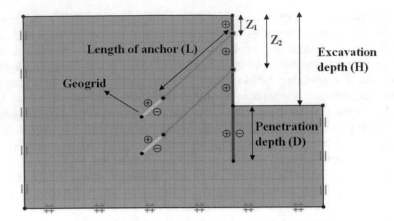

Fig. 1. A typical mesh of anchored sheet pile wall

lower bound elements. The variation of the factor of safety with the total number of elements in the mesh is shown in Fig. 2. It is noted that with an increase in the total number of elements more than 8000 in the mesh, change in the factors of safety is very marginal and insignificant. Moreover, with the change in the total number of elements in the mesh from 8000 to 12000, computational time increases very rapidly. So, for all the parametric study executed in the present study, 8000 elements are reasonably selected for the computational analysis.

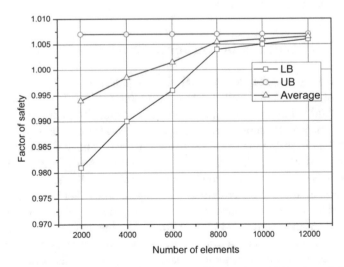

Fig. 2. The variation of the factor of safety corresponding to the total number of elements in the mesh

3   Results and Analysis

First of all, a numerical study is conducted for the evaluation of the required depth of embedded sheet pile (depth of sheet pile below the excavated ground) for a stable sheet pile wall system for the height of excavation (H) of 10 m in the present study. The required embedded depth of sheet pile wall (D) is investigated for a stable system, which is corresponding to the minimum embedded depth of sheet pile wall required to achieve a factor of safety equals to one by varying the depth of embedded wall from 2 m to 10 m i.e. the normalized depth (D/H) of embedded sheet pile ranging from 0.2 to 1.0. The variation of the factor of safety in upper and lower bound are assessed for D/H ranging from 0.2 to 1 and the same is shown in Fig. 3. It is observed that once the normalized depth of penetration the sheet pile wall is increased from 0.2 to 1.0, the factor of safety increases upto D/H = 0.6 (i.e. 6 m) and achieved a factor of safety just greater than one. It has been noted that a further increase in D/H = 0.6 to 1.0, the factor of safety improves marginally. Based on the results obtained from the aforementioned analysis, the depth of embedded sheet pile was fixed as 6 m for the execution of the further parametric study.

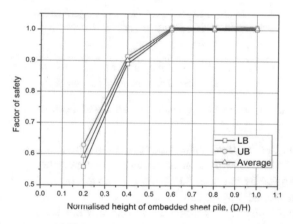

Fig. 3. Variation of the factor of safety with the height of embedded depth of sheet pile

   Provision of anchors to sheet pile wall improves the overall stability of the anchored sheet pile wall system, however, there are many parameters regarding the anchors, which greatly influence the performance of anchored sheet pile wall system. All the important parameters which influence the overall stability of the anchored sheet pile wall system are investigated and discussed in the further section.

   To investigate the effect of variation of the position of anchor measured from the ground surface (z), the normalized depth of anchored located (z/H) is varied from 0 to 1.0 among the sheet pile wall and factor of safety with the consideration of lower and upper bound elements are evaluated. For this analysis, an anchor of a sufficient length of 10 m having a diameter of 0.2 m is considered for the present study. However, it is common to provide the geogrid at the end of the anchor, so a 3 m length of geogrid is attached at the end of anchor throughout the present study. It is noted from the obtained variation of the factor of safety with respect to the position of anchor as shown in Fig. 4, that the FOS of the system gets maximized at two positions of the anchor i.e. z/H = 0.2 and 0.8. The plot showing the variation of a factor of safety with respect to the anchor position in Fig. 4, reveals that either anchor should be placed near the ground surface at z/H = 0.2 or the near the excavation bed, z/H = 0.8. However, for further analysis, the anchor position is fixed at z/H = 0.2.

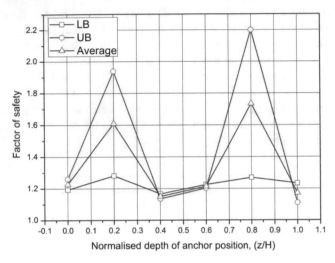

Fig. 4. Variation of the factor of safety with the normalised depth of single anchored position

To economize the length of anchor required to maximize the factor of safety of the overall system, further investigation is carried to evaluate the optimum length of an anchorage system (L) by varying the normalized length of anchor (L/H) ranging from 0.2 to 1.4. For this numerical simulation, the position of the anchor is fixed at z/H = 0.2 based on the results obtained as discussed in the previous section.

A curve showing the variation of a factor of safety with respect to the change in the length of an anchor with consideration of lower and upper bound elements is shown in Fig. 5. From the aforementioned curve, it is observed that the FOS of the system increases with increase in L/H and reaches a maximum L/H = 1.0 (i.e. L = 10 m). With further increase in the anchorage length, a significant reduction in the factor of safety is observed. So, the length of an anchor having length 10 m (L/H = 1.0) is selected as the optimum choice for anchor length to be provided with sheet pile as a support system for the given excavation system. To look into the mechanism of failure (shear dissipation) of the anchored sheet pile system, captured failure pattern corresponding to all studied lengths of the anchor are shown and compared in Fig. 6. It is observed that for the anchor having shorter L/H ratio ranging from 0.2–0.8, the anchor length is completely enclosed in the generated failure surface (plastic region) including the geosynthetics portion attached at the end of the anchor. As the length of anchor increases (L/H ranging 1.0–1.4), a portion of the anchor's length unaffected by the generated failure surface (enclosed in the elastic reason of the soil). Moreover, the grouted portion of the anchor is also unaffected by the generated failure surface, which subsequently gives strength to the anchor to hold it in its original position. This particular phenomenon holds well in the case of L/H $\geq$ 1.0, however, with further increase in L/H, the FOS of the system decreases. Based on the above observation, anchor having a length of 10 m is selected as the optimum length of anchor for the excavation system under the examination.

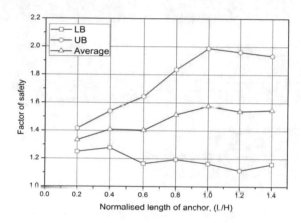

Fig. 5. Variation of the factor of safety with the normalised length (L/H) of the anchor

In practice, anchors to the sheet pile are installed at some inclination with the ground surface (θ). To investigate the effect of inclination of an anchor on the overall stability of the structure, the orientation of anchor (θ) is varied from 15° to 75° having an interval of 10°, while keeping the $z_1/H = 0.2$ and L/H = 1.0. From the numerical simulation where the angle of inclination of anchorage is widely varied, a variation of factor of safety with the inclination angle is shown in Fig. 7. From the analysis, it is observed that the factor of safety increases with θ until a value of 45° and further reduces as the inclination angle of anchor increases. Based on the above observation, the optimum angle of inclination is chosen as 45°. A comparison of failure surface generated by the anchored sheet pile wall having $\theta = 15°, 25°, 35°, 45°, 55°, 65°, 75°$ is shown in Fig. 8. It is worthy to note here that the failure mechanism of the sheet pile wall systems is greatly controlled by the inclination angle of an anchor.

From the Fig. 8, it is noted that in case of $\theta = 45°$, a complex failure surface comprise of two distinctive failure planes is generated and maximum shear dissipation is observed along the length of the anchor. However, in other cases of $\theta \neq 45°$, a clear triangular wedge-shaped failure surfaces are observed and maximum shear dissipation is observed as generated from the base of sheet pile wall and subsequently moves along the sheet pile except for $\theta = 75°$, where a large wedge entirely participates in shear dissipation. Moreover, the observed failure surface is found to be the widest in the case of $\theta = 45°$, which reflects the fact that the maximum soil mass participates in the shear dissipation compared to all other cases. The aforementioned phenomenon contributed towards the higher stability of the overall system i.e. the maximum factor of safety in case of the anchor is inclined at 45°.

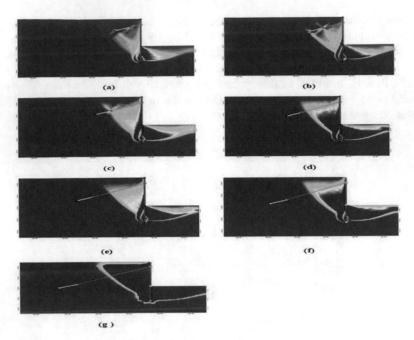

Fig. 6. Failure pattern of normalised anchored sheet pile wall with normalised length of anchor (a) L/H = 0.2 m; (b) L/H = 0.4 m; (c) L/H = 0.6 m; (d) L/H = 0.8 m; (e) L/H = 1.0 m; (f) L/H = 1.2 m; and (g) L/H = 1.4 m

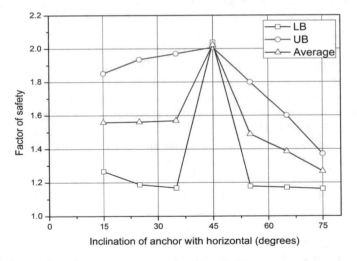

Fig. 7. Variation of the factor of safety with the inclination of an anchor with horizontal

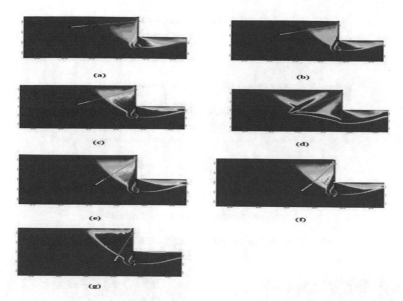

Fig. 8. Comparison of failure surface generated with the variation of inclination of anchor (a) θ = 15° (b) θ = 25° (c) θ = 35° (d) θ = 45° (e) θ = 55° (f) θ = 65° and (g) θ = 75°

To support a high excavation, it is common to use more than one anchor per meter length of the sheet pile wall along with the height of the sheet pile wall. In view of above, to investigate the effect of more than one anchors on the behavior of sheet pile, a study is carried out to investigate the position of the second anchorage system while keeping the first anchor fixed at z/H = 0.2 with L/H = 1.0 and θ = 45°. The position of second anchor is varied at z/H = 0, 0.4, 0.6, 0.8 and 1.0 along with the height of wall with parameters L/H = 1.0 and θ = 45°. A plot for the variation of a factor of safety of the anchored sheet pile wall system with the change in the position of the anchor is shown in Fig. 9.

It can be noted from Fig. 9 that the behavior of the sheet pile wall with second anchor provisioned at different wall heights will affect the overall factor of safety. Based on the variation of FOS with z/H of the second anchor, it is found that anchor positioned at z/H = 0.6 is the best location to provide the maximum FOS with the two anchors for the studies sheet pile wall system. A comparison of failure surfaces obtained from the sheet pile wall with double anchors in shown and compared in Fig. 10. Further, it is worth to mention here that with the provision of second anchor, a factor of safety increases considerably in range of 59% compared to the case where single anchorage was provided with an optimized configuration corresponding to maximum FOS.

An economic sheet pile wall system is one which has the least wall section with the minimum depth of the embedment below the excavation level. As the anchorage is provided to the sheet pile wall system, factor of safety increases, however, in order to have a comparison with a case where no anchorage was provided and sheet pile wall structure was found just stable (FOS = 1) based on sufficient embedded depth of sheet

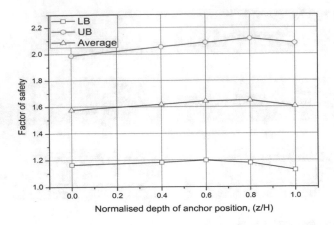

Fig. 9. Variation of factor of safety with the position

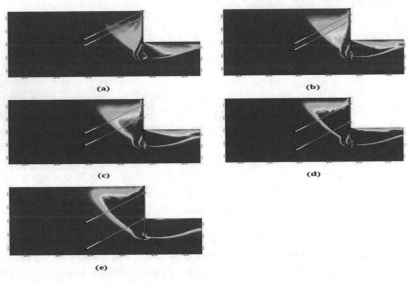

Fig. 10. Failure surface with two anchors having second anchor placed at (a) $z_2 = 0$; (b) $z_2 = 4$ m; (c) $z_2 = 6$ m; (d) $z_2 = 8$ m; (e) $z_2 = 10$ m

pile wall with double anchored sheet pile wall, further study has been carried out. In this study, the embedded depth of double anchored sheet pile wall system having anchors with $L/H = 1.0$ fixed at $z/H = 0.2$ and 0.6 and inclined at $\theta = 45°$. In order to investigate the aforementioned idea, depth of penetration is reduced from 6 m to 2.5 m with an interval of 0.5 m to have a factor of safety just equals to one for the anchored sheet pile wall system. It is self-explanatory that with a decrease in depth of the embedment of sheet pile wall, FOS of overall sheet pile wall decreases, which is similar to the obtained curve from the variation of factor of safety with the decrease in D for the anchored sheet pile wall system as shown in Fig. 11, and a stable sheet pile wall system

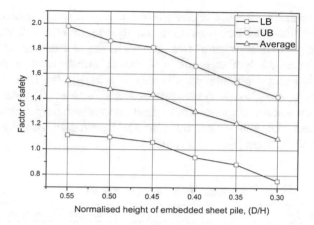

Fig. 11. Variation of the factor of safety with the normalized height of embedded sheet pile

is obtained with D = 3 m with FOS just more than one. With a further decrease in D, anchored sheet pile wall system was found to be unstable. A comparison of failure surface obtained by parametric study with the decrease in D is compared and shown in Fig. 12.

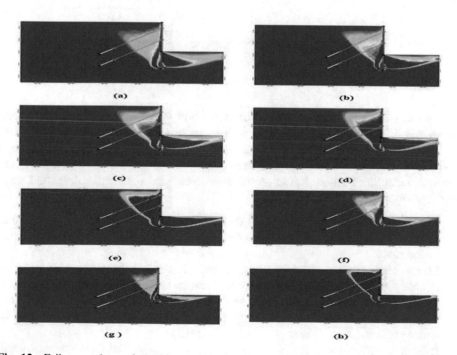

Fig. 12. Failure surface of double anchor sheet pile wall with embedded depth of pile (a) D = 6 m; (b) D = 5.5 m; (c) D = 5 m; (d) D = 4.5 m; (e) D = 4 m; (f) D = 3.5 m; (g) D = 3.0 m; and (h) D = 2.5 m

From the results of the study conducted in the present work, embedment depth of sheet pile wall without anchor system corresponding to the FOS = 1, was found to be 6 m. However, with an inclusion of two anchors having a length of 10 m, it is found that the embedment depth of sheet pile wall reduces substantially, which is equal to 3 m. This embedment depth of sheet pile wall was found 50% lesser in comparison with a case where no anchorage system was provided to sheet pile wall. This outcome reflects that the use of anchorage system to sheet pile wall not only enhances the overall stability of sheet pile wall system but also great saving of steel by reducing the embedment depth of sheet pile wall (Table 1).

Table 1. Material properties

Properties	Soil	Plate (P800)	Anchor plate
Dry unit weight (kN/m^3)	16	–	–
Friction angle (degrees)	35	–	–
Saturated unit weight (kN/m^3)	20	–	–
Young's modulus, E (kN/mm^2)	0.35×10^5	–	–
Poisson's ratio, υ	0.25	–	–
Cohesion, c (kN/m^2)	0	–	–
Normal stiffness, EA (kN)	–	4×10^6	0.42×10^6
Bending stiffness, EI (kN-m)	–	1×10^5	–
Yield force, n_p (kN/m)	–	5000	1000
Yield moment, m_p (kNm/m)	–	800	–

4   Conclusions

The parametric study using finite element method was performed to examine the behaviour of anchored sheet pile wall with the consideration of parameters enlist as embedment depth of sheet pile, anchor's position, length, and inclination with ground and factor of safety of the anchored sheet pile wall system were assessed using the strength reduction method. Based on the aforementioned parametric study, the main findings from the study are as follows.

1. The normalized embedment depth of sheet wall equals to 0.6 was found sufficient to provide a stable sheet pile wall system to support the excavation system considered in the study.
2. In case of sheet pile wall with a single anchor, it is found that the maximum factor of safety of the excavation support system can be achieved by placing anchor near the top or bottom of the excavation. For the studied case, z/H = 0.2 and z/H = 0.8 was found a good choice for placing the anchor, however, one should provide an anchor near the top of sheet pile wall (z/H = 0.2).
3. It is observed that the length of anchor (L) has a great influence on the overall stability of anchored sheet pile wall system and L/H = 1.0 was found the optimum choice for the selection of the length of the anchor.

4. The inclination of anchor measured with horizontal (θ) plays a very crucial role for determining the factor of safety of anchored sheet pile wall system. The best angle for the provision of an anchor was found to be 45°.
5. In order to enhance the factor of safety of anchored sheet pile wall system provision of the second anchor ($z_2 = 6$ m) implement the maximum factor of safety was achieved to corresponding stability of the structure.
6. It is the variation of depth of penetration (6 to 2.5 m) strength parameters at upper and lower bounds then taken the optimum average factor of safety. There will be an optimum factor of safety will be obtained at a 3 m depth of penetration. Its reduces the 50% of steel is used in a sheet pile wall system.

References

OptumG2: Optum Computational Engineering, Copenhagen NV, Denmark (2017)

Bilgin, Ö.: The behaviour of anchored sheet pile walls constructed by excavation and Proceedings of the International Foundation Congress and Equipment Expo, ASCE, Orlando, FL, pp. 137–144 (1994). https://doi.org/10.1061/41023(337)18

Bilgin, Ö., Erten, M.B.: Analysis of anchored sheet pile wall deformations backfilling. MS thesis, School of Civil and Environmental Engineering, Oklahoma State University, Stillwater, OK (2009)

Dawkins, W.P.: Investigation of wall friction, surcharge loads, moment reduction curves for anchored sheet pile walls. US army corps of engineers, ERDC/ITL TR-01-4 (2001)

Bilgin, Ö.: Numerical studies of anchored sheet pile wall behaviour constructed in cut and fill condition. J. Comput. Geotech. **37**(3), 399–407 (2010). https://doi.org/10.1016/j.compgeo.2010.01.002

Performance of Geosynthetic Reinforced Segmental Retaining Walls

Ratnesh Ojha and Vinay Bhushan Chauhan[✉]

Civil Engineering Department, Madan Mohan Malaviya University
of Technology, Gorakhpur 273010, India
ratneshojha96@gmail.com,
chauhan.vinaybhushan@gmail.com

Abstract. Soil reinforcement using Geogrid in the retaining wall is used to strengthen the overall structure to improve the stability of the retaining wall system as well as the behavior of the structure under the seismic loading. With reference to the construction of a conventional rigid retaining wall, the provision of geogrid reinforcement is much appreciated in terms of the performance of the overall structure. Modular (segmental) blocks are used for the construction of such geosynthetic-reinforced soil retaining walls. Such structures offer a wide range of aesthetic finishes, while resulting in more economical structures compared to conventional rigid retaining walls. The present study discusses the effect of the provision of geogrid length in multiple layers on the stability of reinforced retaining wall using numerical modeling under the gravity and seismic loading conditions. This study further investigates the development of a number of failure modes of reinforced soil wall with varying geogrid length configuration. It is established that variation in the length of geogrid not only increases the factor of safety of the wall system but also modifies the failure mode of the reinforced soil retaining wall.

Keywords: Numerical modeling · Geosynthetics · Strength reduction method · Segmental retaining wall

1 Introduction

A retaining wall is a structure used to resist the lateral earth pressure exerted by the material retained behind the wall. Such structures are an essential part of almost all infrastructure projects, to support vertical or near-vertical backfills. Although a variety of retaining walls are in practice to retain the backfill. However, in few past decades reinforced soil walls are being adopted in the majority of projects to provide a sustainable solution for the wall. Soil reinforcement is a method to enhance the stability of overall structure by using a tensile reinforcement in the form of geosynthetics. From the previous studies, it is found that reinforced soil retaining wall performs better under seismic load compared to the convention rigid retaining walls (Ling et al. 2012). Such mechanically stabilized earth walls perform well in case of seismic loading due to their inherent flexibility. It is also noted such walls have been successfully used in practice for transportation and other applications since the 1960s. Owing to the aforementioned

© Springer Nature Switzerland AG 2020
H. Shehata et al. (Eds.): GeoMEast 2019, SUCI, pp. 196–206, 2020.
https://doi.org/10.1007/978-3-030-34184-8_13

structural aspect, easy construction and the cost-effectiveness, reinforced soil retaining wall are leading to the adoption of such structure for retaining the soil system (Yoo and Jung 2006; Liu and Won 2009). These type of wall facing can be geosynthetic-wrapped around, stone-filled gabion baskets, modular blocks, and concrete panels (Han 2015). Modular (segmental) blocks are being widely used for the construction of geosynthetic-reinforced soil retaining walls. Such structures offer a wide range of aesthetic finishes while resulting in more economical structures compared to conventional rigid retaining walls (Huang et al. 2009). Modular-block (segmental) units are typically connected to geosynthetic reinforcement layers using mechanical and/or frictional connection devices. For the construction of such reinforced earth walls, a modular concrete block is laid first along the length of wall and soil is laid behind the modular block and compacted upto the height of the already laid modular block. Before piling up another concrete block over the first one, a geogrid layer of the desired width is laid over the backfill along the length of the wall. This layer is clamped between the two successive modular block layers along with the height of the wall. Once the reinforcement layer is laid, backfill material is filled upto the top level of modular brick and compacted is well compacted. This procedure is continued till the desired height of wall is attained. The deadweight of the modular blocks contributes to the global stability of the geosynthetically reinforced soil retaining wall. The contribution of the facing was neglected in early designs, however, in the limit equilibrium design analysis, the weight of modular blocks was accounted for the analysis (Leshchinsky 2004).

It has been noted from the previous study that due consideration has not been paid towards the numerical simulation of the modular geogrid reinforced wall. So in this present study, a systematic numerical analysis is carried out to examine the behavior of the reinforced soil retaining wall with consideration of varying parameters such as the effect of length of geogrid and horizontal seismic acceleration. This analysis has been carried out using computation numerical tool based on finite element analysis using Optum-G2 to assess the stability of the retaining wall by strength reduction method for reinforced soil retaining wall under the application of seismic loading for varying geogrid length and the performance of reinforced soil retaining wall under seismic conditions has been evaluated. Furthermore, this analysis also examines the effect of length of geogrid on the failure pattern of the reinforced soil retaining wall under static and seismic condition.

2 Numerical Modelling of Geosynthetic Reinforced Wall

In the present study, to carry out the numerical analysis of geogrid-reinforced soil retaining wall, a series of organized simulations are executed using a finite element based computational tool Optum G2 in two-dimensional analysis. The program carries out plane strain analyses for non-seismic as well as the seismic case. For the seismic case pseudo-static analysis is adopted for the present study. The stability of the structure is measured using the strength reduction method with the consideration of reduced strength in solids approach and factor of safety are calculated in the upper and lower bound analysis.

The present paper describes a study of a geogrid reinforced soil retaining wall constructed with a concrete-blocks as a wall to access the stability of such walls in gravity as well as seismic loading. The numerical modeling technique used in the present study majorly focuses on the inclusion of modular-block interaction with backfill soil as well as the reinforcement layer.

A typical mesh considered for this analysis having complete details of the geometrical configuration of geogrid reinforced soil retaining wall having a height (H) of 8 m is shown in Fig. 1. Geogrid material is taken as an elastic material to counteract the tensile stresses developed in the geogrid. The wall is constructed in layerwise having a lift of 1 m (H/8) high soil depth in each construction stage and geogrid is overlaid on the compacted backfill surface and clamped between two successive modular blocks having the size of 1 m high and 2 m wide.

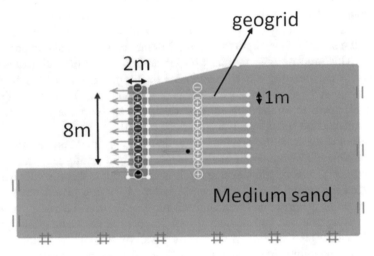

Fig. 1. Typical mesh for the geogrid reinforced soil retaining wall considered in the present study

In the present analysis, modular concrete blocks are considered as rigid material having a unit weight of 23 kN/m^3. Concrete block-geogrid and soil-block interactions were simulated using interface elements using the interface reduction factor of 0.85. As in the practice, each backfill layer is compacted before placing another geogrid layer so that medium soil strata may be achieved. Considering the above fact, the medium sandy soil having Mohr-Coulomb failure criterion is considered as the backfill material in the present study. The Mohr-Coulomb failure criterion is modeled with consideration of non-associated flow rule having a dilation angle of 5°. Material properties used in the present study are summarised in Table 1.

To provide the adequate boundary conditions for the reinforced soil wall system, the bottom is fixed and roller supports were provided at the vertical edges of the mesh. For the simulation of the mesh mentioned above, the total number of ten thousand elements are used in the mesh are generated for the reinforced soil retaining wall system.

Table 1. Material properties considered in the present study

Property	Geogrid	Backfill material
Stiffness (kN/m)	450	–
Yield force (kN/m)	45	–
Modulus of elasticity (MPa)	–	35
Poisson's ratio	–	0.25
Cohesion (kN/m^2)	–	0
Dry unit weight (kN/m^3)	–	18
Saturated unit weight (kN/m^3)	–	20
Internal friction angle (degrees)	–	35
Dilation angle (degrees)	–	5

It has been noted from the previous studies that the length of geogrid provided in the reinforced soil retaining wall has a great influence on the stability of such wall (Liu and Won 2009; Ling et al. 2012). The provision of geogrid reinforcement for the construction of wall not only enhances the stability of the reinforced retaining wall but also the failure mode of the reinforced retaining wall is controlled by the configuration of geogrid layers provided in the retaining wall system. In order to investigate the influence of geogrid length on the stability of the retaining wall system has been studied by varying the length of the geogrid reinforcement. In the present study, the length of reinforcement is varied from 1 m to 20 m at an interval of 1 m i.e. the length of geogrid (L) is varied from 1–20 m. Furthermore, stability assessment of reinforced retaining wall under seismic loading using pseudo-static method is carried out for a range of horizontal seismic acceleration coefficient, k_h = 0.1, 0.3 and 0.5 and compared with the non-seismic case (k_h = 0). Failure modes of the wall and their transition from one mode to another mode of failure due to geogrid length configuration has been studied and the results obtained from the analysis has been discussed in the next section.

3   Result and Discussion

A parametric study was carried out by varying the geogrid length to investigate the effect of length of geogrid on the overall factor of safety of the reinforced soil wall. Variation of the factor of safety versus geogrid length with error bars indicating the worst-case error between upper and lower bound solutions for non-seismic cases (k_h = 0) is shown in Fig. 2. From the plot shown in Fig. 2, it is observed that with the inclusion of geogrid layer, the factor of safety of reinforced soil retaining wall improved substantially, which is obvious as greater the reinforced length of geogrid, more is the resistance force offered by the geogrid layer. It is worth to note from Fig. 2, with an increase in the length of geogrid, the factor of safety gets improved substantially up to L = 12 m. A substantial improvement in the factor of safety was found in the range from 1.0 to 1.83 which is equivalent to 83%. With further increase in the length of geogrid, the factor of safety of the reinforced soil retaining wall system does

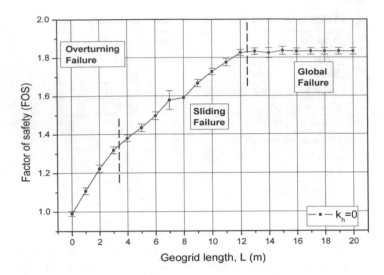

Fig. 2. Variaton of the FOS with geogrid length with error bars indicating the worst-case error between upper and lower bound solutions for non-seismic cases ($k_h = 0$)

not improve, which reveals that optimum geogrid length value is 12 m for the given height of the wall with given geometrical configuration and reinforcement arrangement.

Results obtained from the numerical simulation carried out to assess the seismic stability with the variation of $k_h = 0.1$ to 0.5, are presented in Figs. 3–5. From the plots shown in Figs. 3–5, it is observed the unreinforced retaining walls are unstable for the range of horizontal seismic acceleration considered in the present study. However, with the application of geogrid as a reinforcing layer in the soil wall system, the overall stability of the structure increases. Furthermore, the reinforced soil retaining wall system could achieve an overall factor of safety more than one in case of $k_h = 0.1$ and $k_h = 0.3$. The minimum length of geogrid required to achieve the stability (corresponding to optimum FOS = 1) of reinforced soil retaining wall system increases with increase in k_h i.e. 2 m and 11 m in case of $k_h = 0.1$ and $k_h = 0.3$, respectively (Figs. 3 and 4). Furthermore, in case of $k_h = 0.5$, any length of reinforcement could not achieve provide a stabile reinforced soil retaining wall system with the configuration studies in the present study having spacing between successive geogrid layers equal to H/8 for the given wall, as shown in Fig. 5. This observation highlights the fact that to achieve greater stability of reinforced soil retaining wall at $k_h = 0.5$, the geogrid reinforcement configuration needs to be revised by either reducing the gap between successive geogrid layers or using a geogrid with higher stiffness.

From the Figs. 3–5, it is noted that for all the cases of wall studied in the present study, there exists a maximum length of geogrid layer, as beyond which any further increase in reinforcing length, there is no further increase the factor of safety. In addition, the geogrid length of the reinforcing layer controls the mode of failure of the reinforced soil retaining wall.

A typical comparison of various modes of failure of the reinforced soil retaining wall corresponding to the various length of geogrid for seismic ($k_h = 0.1$; $k_h = 0.3$ and

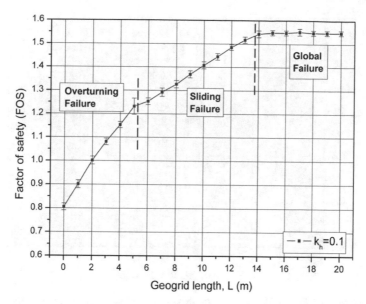

Fig. 3. Variation of FOS with geogrid length with error bars indicating the worst-case error between upper and lower bound solutions for seismic case (k_h – 0.1)

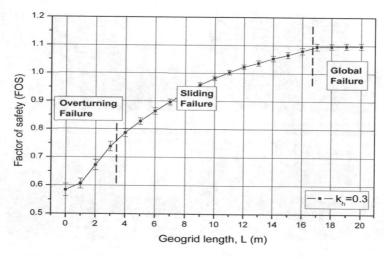

Fig. 4. Variation of FOS with geogrid length with error bars indicating the worst-case error between upper and lower bound solutions for seismic case (k_h = 0.3)

k_h = 0.5) as well as the non-seismic (k_h = 0) cases obtained from the present study are shown in the Figs. 6–11.

In all cases of reinforced wall with non-seismic consideration, overall factor of safety gets improved with increasing length of geogrid length. When the geogrid length is ≤3 m, overturning mode of failure observed and failure surface originates from the

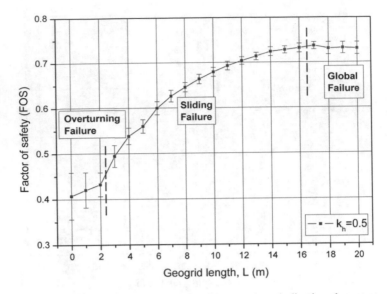

Fig. 5. Variation of FOS with geogrid length with error bars indicating the worst-case error between upper and lower bound solutions for seismic case ($k_h = 0.5$)

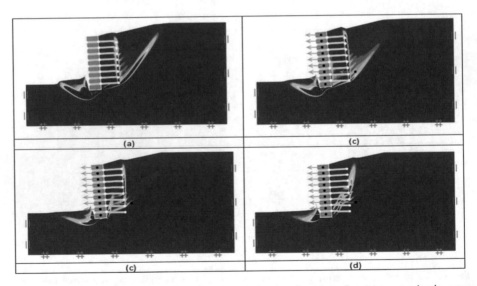

Fig. 6. Mode of failure of reinforced soil wall having L = 3 m for (a) non seismic case; (b) $k_h = 0.1$; (c) $k_h = 0.3$; and (d) $k_h = 0.5$

base of wall and bifurcates at the end of the bottom-most geogrid and it moves upwards along the ends of reinforcing layers as well as into the backfill as shown in Fig. 6(a). As the length of reinforcement increases from 3 m to 12 m, failure surfaces gets distinctive and phenomenon of bifurcation of failure plane reduces with increasing length of

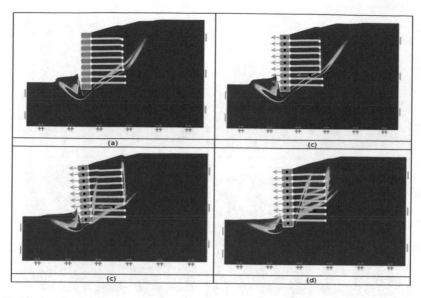

Fig. 7. Mode of failure of reinforced soil wall having L = 6 m for (a) non seismic case; (b) k_h = 0.1; (c) k_h = 0.3; and (d) k_h = 0.5

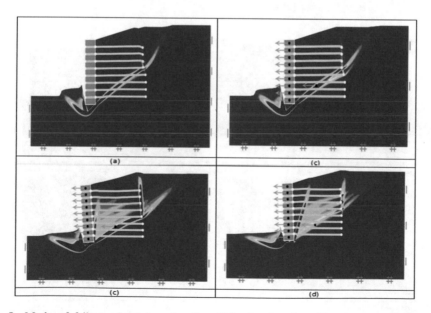

Fig. 8. Mode of failure of reinforced soil wall having L = 9 m for (a) non seismic case; (b) k_h = 0.1; (c) k_h = 0.3; and (d) k_h = 0.5

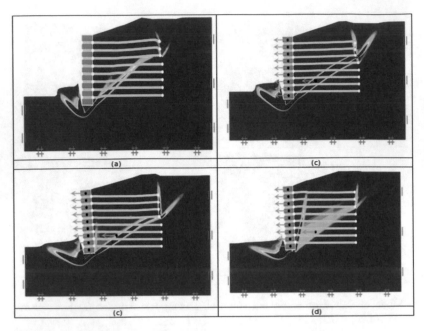

Fig. 9. Mode of failure of reinforced soil wall having L = 12 m for (a) non seismic case; (b) k_h = 0.1; (c) k_h = 0.3; and (d) k_h = 0.5

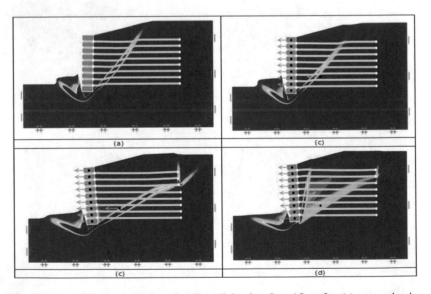

Fig. 10. Mode of failure of reinforced soil wall having L = 15 m for (a) non seismic case; (b) k_h = 0.1; (c) k_h = 0.3; and (d) k_h = 0.5

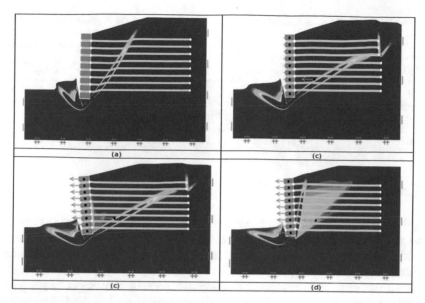

Fig. 11. Mode of failure of reinforced soil wall having L = 18 m for (a) non seismic case; (b) k_h = 0.1; (c) k_h = 0.3; and (d) k_h = 0.5

geogrid, and the sliding failure mode prevails in all such cases as noticeably shown in Figs. 7(a), 8(a) and 9(a). Once the length of geogrid layer increases beyond the 12 m, a clear distinct failure plane prevails and global failure is observed as depicted in Figs. 10(a) and 11(a).

In all cases of reinforced retaining wall with k_h = 0.5, it is worth to note from the results obtained from the present analysis that when the length of reinforcing geogrid is not sufficient to provide a stable reinforced wall system, either two distinct failure planes or a zone of failed backfill is observed based upon the length of reinforcing layer as shown in Figs. 6(d), 7(d), 8(d) and 9(d), 10(d) and 11(d).

In case of reinforced retaining wall having k_h = 0.1 and k_h = 0.3, overturning failure prevails till reinforcing length L $\leq$ 5.25 and L $\leq$ 3.25, respectively as shown in Figs. 6(b, c) and 7(b, c). With an increase in length of geogrid, failure plane gets inclosed well within the reinforced backfill width as depicted in Figs. 8(b, c), 9(b, c), 10(b, c) and 11(b, c).

It is noted that there exists a certain bracket of geogrid length which decide the failure mode of the reinforced retaining wall. It is worth to note here that as the length of geogrid length increases, failure modes of wall takes the transition from overturning failure to sliding failure and further leading to a global failure and such interval of geogrid length is greatly influenced by the magnitude of k_h. Ranges of geogrid length for the different failure modes of reinforced soil retaining wall studied are summarized in the Table 2. Furthermore, as the coefficient of horizontal seismic acceleration increases, the bracketed interval of upper and lower limits gets reduced, owing to the fact that higher the k_h value, lower is the overall factor of safety of a given system.

206   R. Ojha and V. B. Chauhan

Table 2. Range of the length of geogrid (L) for the failure modes of reinforced soil retaining wall

Horizontal seismic acceleration coefficient	Mode of failure of retaining wall		
	Overturning	Sliding	Global
$k_h = 0$	L ≤ 3.25	3.25 < L/H ≤ 12.5	L/H > 12.5
$k_h = 0.1$	L ≤ 5.25	5.25 < L/H ≤ 13.75	L/H > 13.75
$k_h = 0.3$	L ≤ 3.25	3.25 < L/H ≤ 16.75	L/H > 16.75
$k_h = 0.5$	L ≤ 2.33	2.33 < L/H ≤ 16.5	L/H > 16.5

4 Conclusion

The present study involves a finite element numerical analysis to examine the stability of the reinforced soil retaining wall with the modes of failure with the variation of geogrid length. From the results obtained in the study, it has been noted that the provision of geogrid has a prominent effect on the improving the overall stability of reinforced retaining walls compared to unreinforced wall and length of geogrid significantly controls the stability of the overall structure as well as the failure modes of the reinforced soil wall system. For all cases (seismic as well as non-seismic), it has been observed that a factor of safety of wall increases with increase in the geogrid length upto a certain value of the length of geogrid. Further increase in the length of geogrid does not alter the overall factor of safety of reinforced soil retaining wall system. An unstable retaining wall system under seismic loading condition can be configures as a stable with suitable provision of reinforcement in the soil wall system. Moreover, there exist a certain length of the geogrid layer which determined the mode of failure of the reinforced retaining wall system.

References

Han, J.: Principles and Practice of Ground Improvement. Wiley, Hoboken (2015)
Huang, B., Bathust, R.J., Hatami, K.: Numerical study of reinforced soil segmental walls using three different constitutive soil models. J. Geotech. Geoenv. Eng. (2009). https://doi.org/10.1061/ASCEGT.1943-5606.0000092
Leshchinsky, D., Han, J.: Geosynthetic reinforced multitiered walls. J. Geotech. Geoenv. Eng. (2004). https://doi.org/10.1061/ASCE1090-0241(2004)130:12(1225)
Ling, H.I., Leshchinsky, D., Mohri, Y., Wang, J.: Earthquake response of reinforced segmental retaining walls backfilled with substantial percentage of fines. J. Geotech. Geoenv. Eng. (2012). https://doi.org/10.1061/(ASCE)GT.1943-5606.0000669
Liu, H., Won, M.S.: Long-Term Reinforcement Load of Geosynthetic-Reinforced Soil Retaining Walls. J. Geotech. Geoenv. Eng. (2009). https://doi.org/10.1061/(ASCE)GT.1943-56060000052
Yoo, C., Jung, H.Y.: Case history of geosynthetic reinforced segmental retaining wall failure. J. Geotech. Geoen. Eng. (2006). https://doi.org/10.1061/(ASCE)1090-0241(2006)132:12(1538)

Numerical Analysis for the Evaluation of Pull-Out Capacity of Helical Anchors in Sand

Akhil Pandey and Vinay Bhushan Chauhan[✉]

Civil Engineering Department, Madan Mohan Malaviya University
of Technology, Gorakhpur 273010, India
akhilpandey812@gmail.com,
chauhan.vinaybhushan@gmail.com

Abstract. Plate anchors play a significant role in the stabilization of various geotechnical structures. In recent years, the use of helical anchors has extended beyond their traditional use in the foundations. Faster installation and instant loading capabilities have increased their use in multiple infrastructure applications. Through the past few decades, researches have been carried out many studies to determine the pull-out capacity of plate anchor, which has led to the development of many analytical solutions. Unfortunately, the current understanding of helical anchors is unsatisfactory to lay the appropriate framework to be followed by practicing engineers. To have an economic and safer design which leads to increased confidence in design, this study is aimed to use numerical modeling techniques to understand the behavior of helical anchors in the sand and their failure mechanism. Subsequently, the present study focuses on the numerical investigation of pullout capacity of helical anchors with the consideration of various important parameters like embedment depth (H) and anchor plate diameter, D provisioned in the sands of the wide range of relative density i.e. loose, medium and dense sand. It is noted from the results of the present study that the pullout capacity of an anchor provided at sufficiently high value of embedment ratio (H/D) is in excellent agreement with the exact solution proposed by previous researchers. Moreover, a numerical analysis using finite element method with consideration of upper and lower bound limit analysis is carried out for the evaluation of the pull-out capacity of plate anchor (q_u) in multi-layered sand deposit and the variation of q_u with H/D ratio is also presented. Furthermore, the associated failure mechanism of the plate anchors is also analyzed and discussed thoroughly in the present work.

Keywords: Numerical modeling · Helical anchors · Pull-out capacity · Finite element method

1 Introduction

Structures like high rise building, buried pipeline, suspension bridges, transmission towers, retaining wall system, bulkheads, etc. are not only subjected to compression forces but also sometimes experience tensile forces in their life span due to many external causes like weather condition, differential ground settlement or surrounding man-made activities. In such cases, measures like anchors are used to resist the uplift

© Springer Nature Switzerland AG 2020
H. Shehata et al. (Eds.): GeoMEast 2019, SUCI, pp. 207–218, 2020.
https://doi.org/10.1007/978-3-030-34184-8_14

forces and overturning moment. With the advantages of easy installation in any weather condition, decreasing possible damage to the structures from soil movement, generating small or no vibration during installation and easiness in predicting capacity after installation these anchors gain very much popularity in the field of construction.

With the variety of configurations, anchors are manufactured like plate anchors, pile anchors, grouted anchors, prestressed concrete anchors and single and multiple-screw helical anchors (Ghaly et al. 1991). A helical anchor consists of a steel shaft with one or more helical shaped plates welded around it. The central circular solid shaft and helical plates are made up of steel, and the shaft transfers the axial load to the helical plates during the foundation loading (Mosquera et al. 2015).

The uplift capacity of the helical anchor is greatly affected by many factors such as the diameter of the helical anchor plate, embedment depth ratio, etc. Soil characteristics such as density, internal friction angle, etc. also play an important role in defining the maximum pull-out capacity of anchor plate system Niroumand et al. (2012). Therefore, the knowledge of uplift resistance of plate anchors in varieties of soil deposit is necessary where the foundation is required to be designed to withstand the upward or tensile forces acting on it.

It is noted from the literature review that the many previous studies were performed to evaluate the ultimate pull-out capacity of the plate anchors in the case of single-layer sand (Ghaly et al. 1991; Kouzer and Kumar 2009), however, there are limited studies are reported in the available literature for the evaluation of q_u in multi-layered sand deposits (Bhattacharya and Kumar 2016).

In view of above, a comprehensive numerical study based on finite element analysis using OptumG2 has been carried out to evaluate the q_u of the helical anchor system in single layer sand for 3 cases of sand deposits with loose, medium and dense sand. Moreover, q_u of the helical anchor system in double-layered sand deposits analyzed. For both cases of a single and double layer of sand deposit, the variation of q_u of the helical anchor system with H/D ratio is studied and the failure surfaces generated by the various cases are compared and discussed in the present study.

2   Numerical Modelling of Helical Anchor Plate Anchor

Finite element analysis increase accuracy, efficiency, reliability and reduce the uncertainty of the design process Andresen et al. (2011). Hence in this study maximum pull-out capacity of helical anchor plate embedded in horizontally in a different type of sand is analyzed with the upper and lower bond analyses combine with the finite element based computational tool OptumG2.

Furthermore three-dimensional analysis is considered for the analysis of anchor problem, however, the assumption of idealize the anchor as a continuous strip under plane strain condition provide numerical advantages to treat the anchor problem as two dimensions. Hence the two-dimensional helical anchor model is considered for the present study.

A rectangular domain is adopted for this analysis, the depth and width are chosen in a manner that an increment in the size of the domain does not cause any change in the magnitude of the maximum pull-out capacity of the helical anchor system. Appropriate boundary conditions are applied at the edges of the mesh i.e. fixed boundary condition at the bottom of the mesh and roller boundary condition at the vertical ends of the soil mass are chosen. Different sand such as single-layer consist of dense, medium and loose sand and multilayer sand having different depth is used for the embedded helical anchor system following Mohr's failure criteria. As friction force along the failure surface is ignored during Mohr's failure criteria Jinyuan et al. (2012), hence friction force along the failure surface is neglected during the calculation of maximum pull-out capacity. The diameter of the helical plate (D) is fixed at 3 m, very thick helical plate (0.1 m) with respect to the height of the anchor is considered in this study.

The material properties of helical anchor, sand and shaft in a assemble for the numerical simulation are shown in Table 1. In the present study, an axis-symmetry helical model system is preferred thus only half of the model is considered for numerical simulation. There exists symmetry about the central shaft of a helical anchor which helps in easy execution of the numerical simulation. Helical anchor plates are represented by a horizontal line having half diameter (1.5 m), which is connected with the help of connectors to the shaft of the anchor.

Table 1. Material properties

Properties	Sand			Helical anchor plate (P800)	Contral shaft (C1000)
	Loose	Medium	Dense		
Normal stiffness (EA), kN/m	–	–	–	4.0×10^6	4.2×10^5
Bending stiffness (EI), kNm^2/m	–	–	–	1.0×10^5	–
Yield force (n_p), kN/m	–	–	–	5.0×10^3	1.0×10^4
Yield moment (m_p), kNm/m	–	–	–	8.0×10^2	–
Poisson's ratio (v)	0.2	0.25	0.3	–	–
Young modulus (E), MPa	20	35	50	–	–
Cohesion (c), kPa	0	0	0	–	–
Friction angle (ϕ), degrees	30	35	40	–	–

Helical anchor plate and circular shaft are considered as a rigid material. The uniformly distributed load is applied in a stress-controlled manner so that loading was continued until the collapse or the failure of the helical anchor plate system occurred.

Variation of H/D ratio (1 to 6) and its effect on the maximum pull-out capacity of the helical anchor system is analyzed in the single and multilayer sand with the help of finite element combine with a lower and upper bound limit.

The failure mechanism and the associated rupture surface for helical anchor system subjected to pull-out loading are of great importance in the calculation of the maximum pull-out capacity (Ghaly and Clemence 1998). Hence failure interface pattern is also observed in different helical anchor model at various H/D ratio. Furthermore, in the present analysis, installation effect of plate anchor system into the ground and large strain aspects during the pullout of the anchor are neglected in line with Spagnoli et al. (2018).

3 Result and Discussion

Embedment ratio effect the maximum pull-out capacity of helical anchors (Dickin, 1988). Henceforth variation of H/D ratio (1 to 6) of helical anchor plate embedded in loose, medium and dense sand, fixing the diameter of the helical plate as 3 m and its effect on maximum pull-out capacity is examined in this paper (Fig. 1).

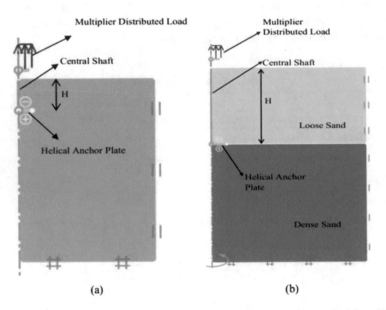

(a)   (b)

Fig. 1. The typical mesh of helical anchor system embedded in (a) sand (b) different layer of sand.

Finite element analysis is performed in the case of helical anchor plate embedded in loose sand and it is found that maximum pull-out capacity increases from 110 ± 1.6 kN/m^2 to 6008.5 ± 41.5 kN/m^2 with increase in H/D ratio from 1 to 6 (shown in Fig. 2). In this study, it is also observed that the percentage in an increment of q_u goes on decreasing with increasing H/D ratio. Variation of H/D from 1 to 2 results in an increment of 290% in q_u, whereas the H/D ratio from 5 to 6 only shows an increment of 60%. The failure mechanism is also observed in this case and it is found that the rupture surface in loose sand originates from the helical plate and propagate towards the ground surface in an upward direction at some angle with vertical. A truncated cone failure pattern is observed which shows a variation in shape along with H/D ratio Niroumand,

H. et al. (2012). It is also found that the size of the zone of influence increases with an increment in the value of H/D (Kumar and Sahoo 2012). At H/D ratio (less than or equal to 3), a narrow shape at the apex of helical plate having small variation in shape in propagating toward the ground surface is observed, while at H/D ratio (greater than 3) widening in shape of failure surface is noticed while propagation of failure surface toward the ground surface (Fig. 3).

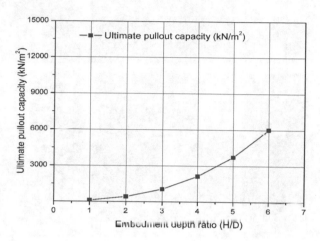

Fig. 2. Variation of q_u with H/D for the helical plate anchor embedded in the loose sand

In medium sand, the effect of the H/D ratio on maximum pull-out capacity is also examined with finite element analysis. Fixing the diameter (3 m) of the helical plate, it is found that q_u increases from 147.45 ± 2.25 kN/m^2 to 9438 ± 61 kN/m^2 with the variation of H/D ratio from 1 to 6 (Fig. 4). It is also noticed that value of q_u is 34% greater than the value which is obtained in the case of helical anchor plate embedded in loose sand at initial stage i.e. H/D = 1 and at H/D ratio equal to 6, q_u of helical plate embedded in medium sand increases to 57%. Although the value of q_u increases with H/D ratio but the percentage increment of q_u goes on decreasing, it decreases from 318% to 62% with the increment of H/D ratio from 1 to 6. In the present study, it is noticed that failure interface show the variation along with the H/D ratio. At H/D (less than or equal to 3) failure pattern demonstrates the narrow zone of influence in the shape of a truncated cone moving vertically toward the ground surface. At H/D ratio (greater than 3) zone of influence of failure surface show the truncated cone with gradually widening of shape with dissemination vertically toward the ground surface (Fig. 5) at some angle with the vertical, effecting large soil masses.

In dense sand, the maximum pull-out capacity of helical anchor plate increases from 247.95 ± 3.65 kN/m^2 to 13525.66 ± 98.45 kN/m^2 at H/D from 1 to 6 (Fig. 6). In this study, it is also found that the maximum ultimate capacity increase rapidly at lower H/D ratio and an increase in the percentage of q_u goes on decreasing at higher H/D ratio. 347.56% increment in q_u is found at change of the H/D ratio from 1 to 2, which falls to 55.68% increment in q_u at H/D ratio from 5 to 6. It also examined that increment in the value of q_u is about 78.23% with respect to helical plate embedded in

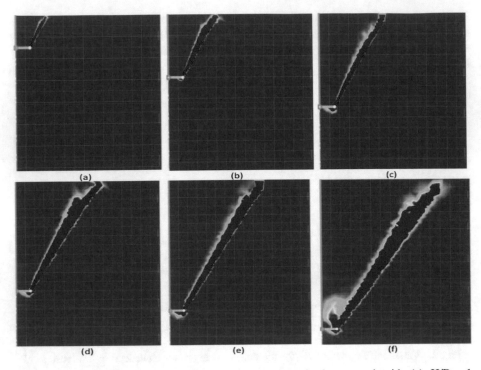

Fig. 3. Failure surface of helical plate anchor systems in loose sand with (a) H/D = 1; (b) H/D = 2; (c) H/D = 3; (d) H/D = 4; (e) H/D = 5; and (f) H/D = 6

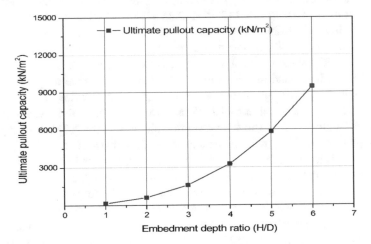

Fig. 4. Variation of q_u with H/D for the helical plate anchor embedded in the medium sand

loose sand and 32.96% with respect to the q_u of helical anchor plate embedded in medium sand at H/D = 1 and at H/D ratio equal to 6, q_u of helical plate embedded in dense increases to 128.78% in comparison with q_u of helical plate embedded in loose sand.

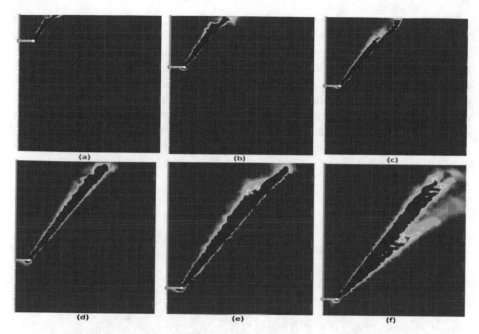

Fig. 5. Failure surface of helical plate anchor systems in medium sand with (a) H/D = 1; (b) H/D = 2; (c) H/D = 3; (d) H/D = 4; (e) H/D = 5; and (f) H/D = 6

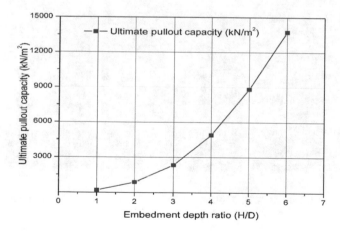

Fig. 6. Variation of q_u with H/D for the helical plate anchor embedded in the dense sand

Failure interface pattern obtained in this case is also examined in the present numerical simulation. In this study it is noticed that failure pattern shows a cone-shaped wedge, shear zones start at the anchor edges and widen progressively until they reach the soil surface Jinyuan et al. (2012) (Fig. 6). At lower H/D ratio (less than or equal to 3) soil mass just above the helical plate is getting effected and slight change is noticed in the cone shape failure pattern. At higher H/D ratio (>3), huge soil mass emanates and flaring of cone shape pattern is noticed at the ground surface (Fig. 7). The pull-out capacity of helical anchor system is greatly influenced with the change in soil characteristics Niroumand et al. (2012) different soil layer can affect the maximum pull-out capacity as well as the failure mechanism of helical plate system. Thus in this study, q_u of helical anchor plate in multilayer sand with different internal friction angle (ϕ) is also analyzed with numerical simulation. Two cases i.e. helical anchor plate embedded in dense sand having $\phi = 40°$ with the loose sand having $\phi = 30°$ just above the helical anchor plate and helical anchor embedded in loose sand and dense sand above the plate is analyzed with the variation of H/D ratio. As the depth of the helical plate increases, the depth of sand above the helical plate is also increasing.

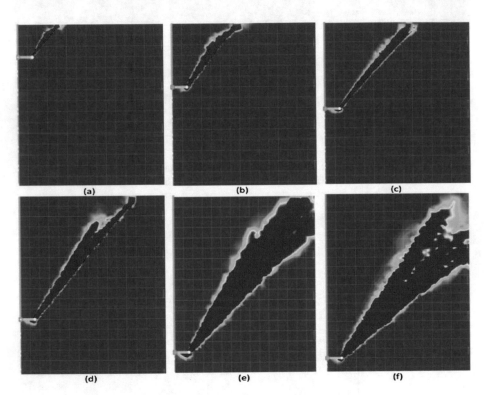

Fig. 7. Failure surface of helical plate anchor systems in dense sand with (a) H/D = 1; (b) H/D = 2; (c) H/D = 3; (d) H/D = 4; (e) H/D = 5; and (f) H/D = 6

It is examined during the finite element numerical simulation that maximum pull-out capacity increases from 247.95 ± 3.65 kN/m^2 to 13525 ± 98.45 kN/m^2 and 441.45 ± 8.25 kN/m^2 to 30876 ± 188.51 kN/m^2 with the variation of H/D ratio from 1 to 6 respectively in both the condition (Figs. 8 and 9). Presence of dense above the helical anchor plate results in an additional increase of 78% in maximum pull-out capacity of helical anchor system at H/D ratio equal to 1 and 128.29% at H/D ratio equal to 6 with respect to helical plate anchor embedded in dense sand having loose

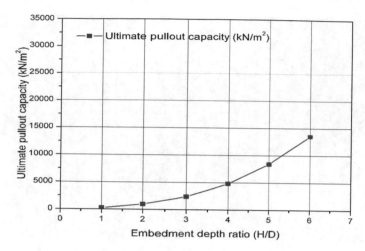

Fig. 8. Variation of q_u with H/D for the helical plate anchor embedded in the dense sand with a layer of loose sand above the helical plate

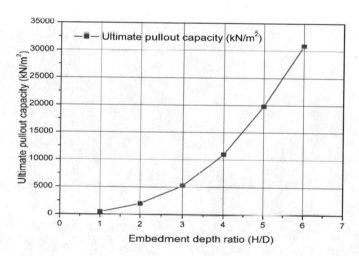

Fig. 9. Variation of q_u with H/D for the helical plate anchor embedded in the loose sand with a layer of dense sand above the helical plate

above the helical plate. Failure interface with rupture of sand shows a variation of shape of truncated cone i.e. widening of rupture surface at higher H/D ratio (greater than 3) in both the cases (Figs. 10 and 11).

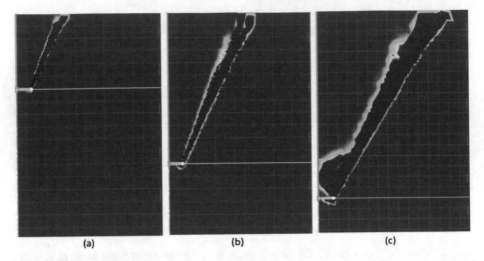

Fig. 10. Failure pattern of helical anchor plate embedded in dense sand and having loose sand above the helical plate with (a) H/D = 2; (b) H/D = 4 and (c) H/D = 6

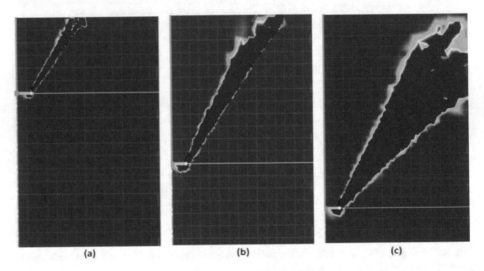

Fig. 11. Failure pattern of helical anchor plate embedded in loose sand and having dense sand above the helical plate with (a) H/D = 2; (b) H/D = 4 and (c) H/D = 6

4 Conclusion

The present study involves comprehensive finite element numerical analysis with a combination of lower and upper bond limit of helical anchor system embedded in loose, medium and dense sand. The effect of H/D ratio on the maximum pull-out capacity of helical anchor plate system in single layer sand (dense, medium and loose) and double layer sand consists of loose sand above the helical plate embedded in dense sand and vice versa is analyzed. The following conclusions are drawn from the numerical simulation of the helical anchor system.

1. In the of case helical anchor plate embedded in single layer sand, q_u of helical anchor plate increases with an increase in H/D ratio embedded in loose, medium and dense sand Increment of q_u is about 290% to 60%, 318% to 62% and 347.56% to 55.68% is observed with the variation of H/D from 1 to 2 and from 5 to 6 in case of helical anchor plate embedded in loose, medium and dense sand respectively. It is examined that maximum pullout capacity increased 34% and 78.23% in medium and dense sand respectively with comparison to q_u of loose sand at H/D ratio = 1, while increment q_u of in helical anchor plate embedded in dense sand with comparison with q_u of helical anchor plate embedded in medium sand is about 32.96% at H/D equal to 1 is observed. When the H/D ratio is equal to 6 then increase in 57% and 128.78% is noticed in medium and dense sand respectively with respect to loose sand.

2. Helical anchor plate system in multilayer sand of varying depth above helical plate, it is found that the value of q_u in the case of dense sand above the helical plate is 78% more than the value q_u at H/D = 1 and 128.29% at H/D = 6 in which loose sand is present just above the helical anchor plate. It is also inspected that at same H/D ratio, q_u in multilayer helical anchor system is greater than q_u in helical anchor plate system embedded in the sand(loose, medium and dense).

3. Failure surface shows the shape of a truncated cone with the ascendant movement of soil masses in the sand, instigate from the helical plate and moves toward the earth surface in loose, medium and dense sand. The cone-shaped failure pattern that observed during this analysis is narrow at its origin point with less change in their shape at H/D ratio less than or equal to 3, but H/D ratio greater than 3 widenings in failure pattern is observed at the ground surface.

References

Andresen, L., et al.: Finite element analyses applied in design of foundations and anchors for offshores structures. Int. J. Geomech. (2011). https://doi.org/10.1061/(ASCE)GM.1943-5622.0000020

Bhattacharya, P., Kumar, J.: Uplift capacity of anchors in layered sand and using finite-element limit analysis: formulation and results. Int. J. Geomech. (2016). https://doi.org/10.1061/(ASCE)GM.1943-56220.0000560

Choudhary, A.K., Dash, S.K.: Load-carrying mechanism of vertical plate anchors in sand. Int. J. Geomech. (2016). https://doi.org/10.1061/(ASCE)GM.1943-5622.0000813

Dickin, E.A.: Uplift behaviour of horizontal anchor plates in sand. Int. J. Geotech. Eng. **114**(11), 1300–1317 (1988)

Ghaly, A.: Uplift behaviour of screw anchors in sand. Int. J. Geotech. Eng. **117**(5), 773–793 (1991)

Ghaly, A.M., Clemence, S.P.: Pullout performance of inclined helical screw anchors in sand. Int. J. Geotech. Geoenvironmental Eng. **124**(7), 617–627 (1998)

Jinyuan, Liu, et al.: Sand deformation around an uplift plate anchor. Int. J. Geotech. Geoenvironmental Eng. (2012). https://doi.org/10.1061/(ASCE)GT.1943-5606.0000633

Kouzer, K.M., Kumar, J.: Vertical uplift capacity of equally spaced horizontal strip anchors in sand. Int. J. Geomech. (2009). https://doi.org/10.1061/(ASCE)1532-3641(2009):5(230)

Kumar, J., Sahoo, J.P.: Upper bound solution for pullout capacity of vertical anchors in sand using finite elements and limit analysis. Int. J. Geomech. (2012). https://doi.org/10.1061/(ASCE)GM.1943-5622.0000160

Mosquera, Z., et al.: Serviceability performance evaluation of helical piles under uplift loading. J. Perform. Constructed Facil. (2015). https://doi.org/10.1061/(ASCE)CF.1943-5509.0000805

Niroumand, H., et al.: Performance of helical anchors in sand. Electron. J. Geotech. Eng. **17**, 2683–2702 (2012)

Spagnoli, G., et al.: Estimation of uplift capacity and installation power of helical piles in sand for offshore structures. J. Waterw. Port Coastal Ocean Eng. (2018). https://doi.org/10.1061/(ASCE)WW.1943-5460.0000471

Author Index

Printed in the United States
By Bookmasters